T. V. NARAYANA is a member of the Department of Mathematics at the University of Alberta.

Lattice path combinatorics has developed greatly as a branch of probability studies recently, and the need for new books on the subject is obvious. The present monograph, by one who has made significant contributions to combinatorics and its applications to probability and statistics, will be useful to research workers, teachers, professional statisticians, and advanced students alike. It treats several recent results and it offers a powerful new tool for studying many problems in mathematical statistics.

The emphasis in the five chapters is on 'dominance.' From a consideration of exceedances in the lattice path problem, the text goes on to provide solutions to tests of hypotheses and simple sampling plans, displaying the usefulness of Young chains in the enumeration of the latter. The fourth chapter, on knock-out tournaments, represents one approach to paired comparisons quite close in spirit to dominance and lattice path combinatorics, and the final chapter considers the advantages of using combinatorial methods in statistical problems (including the Frame-Robinson-Thrall theorem to derive properties of non-parametric tests) and mentions current trends of research. Numerous examples, exercises, and references round out the text.

MATHEMATICAL EXPOSITIONS

Volumes Published

1 *The Foundations of Geometry* G. DE B. ROBINSON
2 *Non-Euclidean Geometry* H. S. M. COXETER
3 *The Theory of Potential and Spherical Harmonics* W. J. STERNBERG and T. L. SMITH
4 *The Variational Principles of Mechanics* CORNELIUS LANCZOS
5 *Tensor Calculus* J. L. SYNGE and A. E. SCHILD (out of print)
6 *The Theory of Functions of a Real Variable* R. L. JEFFERY (out of print)
7 *General Topology* WACLAW SIERPINSKI
(translated by C. CECILIA KRIEGER) (out of print)
8 *Bernstein Polynomials* G. G. LORENTZ (out of print)
9 *Partial Differential Equations* G. F. D. DUFF
10 *Variational Methods for Eigenvalue Problems* S. H. GOULD
11 *Differential Geometry* ERWIN KREYSZIG (out of print)
12 *Representation Theory of the Symmetric Group* G. DE B. ROBINSON
13 *Geometry of Complex Numbers* HANS SCHWERDTFEGER
14 *Rings and Radicals* N. J. DIVINSKY
15 *Connectivity in Graphs* W. T. TUTTE
16 *Introduction to Differential Geometry and Riemannian Geometry* ERWIN KREYSZIG
17 *Mathematical Theory of Dislocations and Fracture* R. W. LARDNER
18 *n-gons* FRIEDRICH BACHMANN and ECKART SCHMIDT
(translated by CYRIL W. L. GARNER)
19 *Weighing Evidence in Language and Literature: A Statistical Approach*
BARRON BRAINERD
20 *Rudiments of Plane Affine Geometry* P. SCHERK and R. LINGENBERG
21 *The Collected Papers of Alfred Young, 1873–1940*
22 *From Physical Concept to Mathematical Structure. An Introduction to Theoretical Physics* LYNN E. H. TRAINOR and MARK B. WISE
23 *Lattice Path Combinatorics with Statistical Applications* T. V. NARAYANA

MATHEMATICAL EXPOSITIONS NO. 23

Lattice path combinatorics with statistical applications

T. V. NARAYANA

UNIVERSITY OF TORONTO PRESS
Toronto Buffalo London

Toronto Buffalo London
Printed in Great Britain

Library of Congress Cataloging in Publication Data

Narayana, Tadepalli Venkata, 1930–
Lattice path combinatorics, with statistical applications.

(Mathematical expositions; no. 23)
Bibliography: p.
Includes index.
1. Mathematical statistics. 2. Combinatorial analysis. 3. Lattice theory.
I. Title. II. Series.
QA276.N33 519.5 78–6710
ISBN 0-8020-5405-6

The first part of this manuscript is dedicated to J. R. McGregor; the latter part (tournaments) to Professor H. E. Gunning, now president of the University of Alberta, as a token of respect for his human qualities and admiration for his scientific achievements.

तं वीक्ष्य वेपथुमती सरसाङ्गयष्टि—
  निक्षेपणाय पदमुद्धृतमुद्वहन्ती ।
मार्गाचलव्यतिकराकुलितेव सिन्धुः
  शैलाधिराजतनया न ययौ न तस्थौ ॥ ८५ ॥

Contents

Tables

Preface

Although the work of Alfred Young (1873–1940) was continued by many algebraists, including G. de B. Robinson, the main development of the combinatorial aspects of Young's work seems to have taken place well after 1950. As this year saw the publication of Volume I of W. Feller's *An Introduction to Probability Theory and Its Applications*, it became possible to study in a unified way not only Young chains and lattice path combinatorics but also their statistical applications.

I have limited my account to problems which can be conveniently treated in terms of 'dominance,' which is a special case of majorization of Hardy, Littlewood, and Polya involving integers only. In the first three chapters I have attempted to motivate both the applications and principal theorems connected with dominance by informal discussion, examples, and many exercises. Chapter IV represents one approach to paired comparisons quite close in spirit to dominance and lattice path combinatorics. It is hoped that the presentation is elementary enough for even an interested undergraduate to understand most of Chapter V and at least glimpse the possibilities of practical applications. I conclude with an exercise explicitly obtaining the basic (or irreducible?) non-parametric tests using the *théorème d'équerre* of Frame, Robinson, and Thrall.

Apart from random remarks on the development of the subject, the section 'Notes and Solutions' gives references to the pertinent literature for solving almost all the easy exercises. The bibliography supplements the references at the ends of the chapters and is meant to be complete only from the point of view of dominance. Also, those exercises which either appear to me difficult or have not been published as far as I know are marked with a dagger (†).

As the notations used are quite standard, only a brief comment is made about certain innovations and abbreviations which might puzzle the reader. I have not hesitated to use two notations for one entity, such as both $N(S)$ and $|S|$ to denote the number of elements in a set, or $\bar{S}$ and S^c for the complement of a set. More questionable might be the abbreviations s.p. for 'sampling plan' and s.ps. for 'sampling plans.' As the acceptance of the combinatorial theory of Young chains into mathematics (not to speak of statistics) has been long overdue in the United States and therefore Canada, I felt justified in attempting to achieve maximum clarity and simplicity by using the abbreviated forms. (After nearly a decade has

Kreweras's Theorem been mentioned in *Mathematical Reviews*? Will Chapter 12 of the recent translation of Kagan, Linnik, and Rao's book [*Characterization Problems in Mathematical Statistics*; John Wiley and Sons, New York, 1973] succeed in focusing attention on Young chains where others have failed?)

Finally, I hope to steer clear of statistical controversies, particularly in using randomized tests to attain the exact levels in the context of non-parametric tests. To some statisticians such randomization is necessary; they should be pleased with the calculations involving exact levels as in Table 5 and in certain exercises. However, other statisticians consider using such randomization (in the same narrow context of exact levels) as worse than pollution: to these we offer the combinatorial methods developed in this book, following the lead of W. Feller as summarized in L. Comtet's *Advanced Combinatorics*. The reader may perhaps agree with me that statistical controversies are irrelevant for dominance (or Young) tests where weak inadmissibility seems most important.

Among many others, it is a pleasure to thank Dr L. Comtet and Dr M. S. Rao for helpful comments. For personal inspiration at decisive moments, the late Yu. V. Linnik and L. Moser (in the past) as well as Professor S. Geisser (I hope also in the future) must be mentioned. Finally grants from the National Research Council of Canada over the last few years have facilitated the computations, and a grant from the University of Alberta was provided to assist publication.

From the practical point of view of publication, I am deeply indebted to Professor G. de B. Robinson for so graciously inviting me to present this manuscript for publication in the Mathematical Expositions Series of the University of Toronto Press. Professor I. R. Savage was kind enough to make several helpful suggestions and I also thank Dr S. G. Mohanty for valuable discussions – although we may not always have seen eye to eye! In particular, the rewritten appendix, though my responsibility, is based on a preprint of Dr Mohanty; any good points in it may naturally be credited to him and any shortcomings to me. Last, but not least, I thank Miss L. Ourom of the University of Toronto Press for her careful editing.

T.V.N.
Edmonton, August 1976

LATTICE PATH COMBINATORICS WITH STATISTICAL APPLICATIONS

I

Lattice path problems and vectors of integers

Surge ai mortali per diverse foci
  la lucerna del mondo; ma da quella
  che quattro cerchi giunge con tre croci,
con miglior corso e con migliore stella
  esce congiunta ...

Dante, *Paradiso*

Just as in the seventeenth century when the beginnings of combinatorial analysis coincided with those of probability theory, many pleasing advances in both combinatorial theory and probability theory have been made in the last twenty-five years. As one example, we prove (and slightly extend) a celebrated theorem of Chung and Feller (1949). We also introduce a special case of the definition of majorization – called domination – of Hardy, Littlewood, and Polya (1952), which simplifies many combinatorial problems associated with lattice paths in the plane. We conclude with an introduction to the ballot theorem and related problems, where simple bijections and domination yield explicit results.

1
REPRESENTATION OF SUBSETS OF $\{1, \ldots, N\}$

Given a subset M of the integers $\{1, 2, \ldots, N\}$ containing $m = |M|$ elements, we set $n = |\bar{M}|$ (so that $N = m+n$) and interpret M as follows:

1. To be given M is equivalent to be given the results of a coin-tossing experiment with N tosses, where we use the convention that

$$t \in M \Leftrightarrow t\text{th toss gave head} \qquad (t = 1, \ldots, N).$$

Thus the numbers m, n represent the numbers of heads and tails in N tosses, so that M can be interpreted as a *random walk*.

2. We can represent M by a polygonal line $\mathscr{P}$ joining the origin $O\,(0, 0)$ to the point

B with coordinates (m, n). Unit horizontal steps correspond to elements of M, and unit vertical steps correspond to elements of $\bar{M}$. Such a line $\mathscr{P}$ is called a *minimal lattice path* or simply *path* since there exist no paths of shorter length $(<N)$ joining O to B and whose elements are of unit length linking the lattice points.

3. We may represent M by the vector of non-decreasing integers $A = (a_1, \ldots, a_n)$, where a_i is the minimal distance measured parallel to the x-axis of the point $(m, n-i)$ from the path $\mathscr{P}$ as defined in 2 above. Clearly

[1a] $0 \leq a_1 \leq a_2 \leq \ldots \leq a_n \leq m$;

conversely, to any vector satisfying [1a] corresponds a unique path. Henceforth we shall refer to such a vector as a path, whenever it is convenient.

4. Finally, we can associate with M, $\mathscr{P}$, or A the 'characteristic' composition $C = (c_0, c_1, \ldots, c_m)$, where, referring to [1a], we define

[1b] $c_i =$ number of a_i's which equal i $\quad (i = 0, 1, \ldots, m)$.

Clearly to any path $\mathscr{P}$ or vector A corresponds the unique composition $C = (c_0, c_1, \ldots, c_m)$ with $\sum_{i=0}^{m} c_i = n$, and conversely.

Other interesting representations are, of course, possible; we content ourselves with definitions of a few properties of paths and an example in the special case where $m = n$. In this case the path starts at $(0, 0)$ and ends at the point (n, n) on the diagonal $x = y$.

DEFINITION A (Area under a path) Given a path $\mathscr{P}_0$, with vector $A_0 = (a_1, \ldots, a_n)$ satisfying [1a] above, we define

[1c] $W(A_0) = \sum_{i=1}^{n} a_i$

as the *area under the path.*

DEFINITION B (Exceedance of a path) Given a vector $A_0 = (a_1, \ldots, a_n)$ satisfying [1a], consider the set of inequalities

[1d] $a_j > j-1 \quad (j = 1, \ldots, n)$.

If exactly r of the above inequalities are true for A_0, we say the *exceedance of the path A_0 is r*. It is easy to verify that exceedance can also be defined as the number of true inequalities among

[1e] $c_0 + c_1 + \ldots + c_{j-1} \leq j-1 \quad (j = 1, \ldots, n)$.

$C_0 = (c_0, c_1, \ldots, c_n)$ is the composition corresponding to A_0, and we shall refer to exceedance of a composition as defined by [1e].

Finally, we can generate $n+1$ paths from A_0 as follows. Consider the vectors B_i, where

[1f] $\quad B_i = (a_1 + i, a_2 + i, \ldots, a_n + i) \qquad (i = 0, 1, \ldots, n).$

Reduce the elements of each B_i modulo $n+1$ and rearrange the elements within each B_i so that they are in non-decreasing order. Let us denote these rearranged B_i's by A_i's for each i, and note that $A_0 \equiv B_0$. Then $A_0, A_1, \ldots, A_n$ are *paths generated by* A_0 or *cyclic permutations of* A_0. Indeed, we see easily that, if $C_0 = (c_0, \ldots, c_n)$ (cf. [1b]) is the composition associated with A_0, then the cyclic permutations of C_0 correspond to the vectors $A_1, \ldots, A_n$ generated by A_0.

DEFINITION C (Cyclic permutations of a vector) Given a path A_0, with composition C_0, let us consider the cyclic permutations C_r of C_0, where

[1g] $\quad C_r = (c_{n-r+1}, c_{n-r+2}, \ldots, c_n, c_0, \ldots, c_{n-r}) \qquad (r = 1, \ldots, n).$

The vectors $A_0, A_1, \ldots, A_n$, where A_r is the vector corresponding to C_r in [1g], are called *cyclic permutations* of A_0.

EXAMPLE Let $n = 4$ and $A_0 = (0,0,1,1)$. C_0 is then $(2,2,0,0,0)$. The cyclic permutations of C correspond to the vectors $A_1 = (1,1,2,2)$, $A_2 = (2,2,3,3)$, $A_3 = (3,3,4,4)$, $A_4 = (0,0,4,4)$ generated by A_0. The exceedances of $A_0, A_1, \ldots, A_4$ are $0, 1, 3, 4, 2$, and $W(A_0) = 2$, $W(A_1) = 6$, $W(A_2) = 10$, $W(A_3) = 14$, $W(A_4) = 8$. Letting $r(A)$ denote the exceedance of A, we notice that $r(A_3) > r(A_2) > r(A_4) > r(A_1) > r(A_0)$ and the areas under A_0, A_1, A_2, A_3, A_4 satisfy the same inequality as the $r(A)$'s. We shall thus prove, in general, that if A_0, A_1, ..., A_n are vectors generated by A_0, the path with greater exceedance has also the greater area under it.

2
A REFINEMENT OF THE CHUNG-FELLER THEOREM

Let us consider the $\binom{2n}{n}$ paths from the origin to (n, n). The Chung-Feller Theorem, which was first proved in 1949 and of which a direct inductive proof is provided by Feller (1968), states: 'The number of paths which have exactly k exceedances is independent of k ($k = 0, 1, \ldots, n$), and hence equals $\frac{1}{n+1}\binom{2n}{n}$.' We refine it as follows. The paths $A_0, A_1, \ldots, A_n$ generated by a path A_0 are all distinct, and the areas under them, i.e. the numbers $W(A_i)$, modulo $n+1$ will be a complete set of residue classes modulo $n+1$. Further, if A_i, A_j are two distinct paths from the set $A_0, A_1, \ldots, A_n$, the path with greater area under it has greater exceedance, i.e. $W(A_i) > W(A_j)$ implies $r(A_i) > r(A_j)$. Thus in each set of $n+1$ cyclic permutations of a path $A_0, A_1, \ldots, A_n$ we can order the paths according to the areas under them as $W(A_{i_1}) > W(A_{i_2}) \ldots > W(A_{i_{n+1}})$ where $i_1, i_2, \ldots, i_{n+1}$ is a permutation of $0, 1, \ldots, n$; then automatically $r(A_{i_1}) > r(A_{i_2}) \ldots > r(A_{i_{n+1}})$. As a consequence, in each equivalence class of the $n+1$ cyclic permutations of a path there is one and only one path with a given exceedance k ($k = 0, 1, \ldots, n$). We state a few lemmas which can be established from our definitions.

LEMMA 2A *If $A_0, A_1, \ldots, A_n$ are the cyclic permutations of A, then $W(A_i) \neq W(A_j)$ if $i \neq j$, and hence these $n+1$ vectors are distinct.*

LEMMA 2B *Given any $(n+1)$-composition of n, $C = (c_0, c_1, \ldots, c_n)$, there exists a cyclic permutation of C with zero exceedance.*

This lemma suggests defining the cumulative composition of $C = (c_0, c_1, \ldots, c_n)$ as $D' = (d_0, \ldots, d_n)$, where $d_j = c_0 + \ldots + c_j$ $(j = 1, \ldots, n)$. Clearly $d_n = n$, and noting that

[2a] $0 \leq d_1 \leq d_2 \leq \ldots \leq d_n = n,$

we see that $D = (d_0, \ldots, d_{n-1})$ (i.e. D' with d_n deleted) is, by [1a], a path vector. From our definitions we get the important result

[2b] The inequality $d_{j-1} > j-1$ is true if and only if

$$a_j \leq j-1 \qquad (j = 1, \ldots, n).$$

We verify further that d_j $(j = 0, \ldots, n-1)$ represent the distances measured parallel to the y-axis of the points $(n-1-j, n)$ from the path. From this interpretation of a cumulative composition vector $D = (d_0, d_1, \ldots, d_{n-1})$, $W(D) = \sum_{j=0}^{n-1} d_j$ is the area 'above' the path, and the following lemma is almost evident.

LEMMA 2C *Given a path A, with composition vector C and exceedance r, let $D' = (d_0, d_1, \ldots, d_{n-1}, n)$ be the cumulative composition vector of C. Then the exceedance of $D = (d_0, \ldots, d_{n-1})$ is $n-r$, and $W(D) + W(A) = n^2$.*

Indeed, since the number of true inequalities $d_{j-1} > j-1$ $(j = 0, \ldots, n)$ is the same for D' and D we identify them when considering exceedances and speak indifferently of the exceedance of D or D'. We abbreviate $r(A_i)$ to r_i in Theorem 2A $(i = 0, 1, \ldots, n)$.

THEOREM 2A (Chung-Feller) *Let C_0 be a $(n+1)$-composition of n and $C_1, \ldots, C_n$ its cyclic permutations. Let $A_0, A_1, \ldots, A_n$ be the paths corresponding to*

$C_0, C_1, \ldots, C_n$ respectively. Then $W(A_i) > W(A_j)$ implies $r_i > r_j$ $(i \neq j;\ i, j = 0, 1, \ldots, n)$.

□Proceeding inductively, we assume the theorem true for all $n \leq k$, and let $C_0 = (0, c_1, \ldots, c_{k+1})$ be the given $(k+2)$-composition of $k+1$, which without loss of generality we assume to have zero exceedance using Lemma 2B. Consider the $(k+1)$-composition of k

$$B_0 = (c_1, \ldots, c_{k+1} - 1) = (b_1, \ldots, b_k)$$

and let $B_1, \ldots, B_k$ be its cyclic permutations; further let $E_1, \ldots, E_{k+1}$ be the cumulative compositions of $B_1, \ldots, B_k, B_0$ respectively. Set

$$\text{[2c]} \quad \left.\begin{aligned} K_{r,s} &= b_{k+2-s} + \ldots + b_{k+2-(r+1)} \\ K_{0,r} &= b_{k+2-r} + \ldots + b_{k+1} \end{aligned}\right\} \qquad (r < s;\ r, s = 1, \ldots, k+1)$$

and note that the kth element of E_k is $K_{0,k}$. Evidently $D_1', \ldots, D_{k+1}'$, the cumulative compositions of $C_1, \ldots, C_{k+1}$, can be obtained from $E_1, \ldots, E_{k+1}$ as follows:

(i) For $j = 1, \ldots, k+1$ introduce a new element $K_{0,j}$ in E_j immediately before its jth element $K_{0,j}$ to obtain a vector E_j'. (This corresponds to the element $c_0 = 0$ of C_0 which is suppressed in B_0.)

(ii) Leaving the first $j-1$ elements of E_j' unaltered, increase all other elements of E_j' by unity to obtain D_j' – thus compensating for the fact that $b_{k+1} = c_{k+1} - 1$ $(j = 1, \ldots, k+1)$.

Now

$$W(E_j') = W(E_j) + K_{0,j},$$

and from (ii)

$$W(D_j') = W(E_j) + K_{0,j} + (k+3-j).$$

Clearly $W(E_j) - W(E_{j-1}) = (k+1)b_{k+2-j} - k$ $(j = 2, \ldots, k+1)$. We conclude from the above two equations and $K_{0,s} - K_{0,r} = K_{r,s}$ that for all $r < s$ $(r, s = 1, \ldots, k+1)$,

$$W(E_s) - W(E_r) = (k+1)K_{r,s} - k(s-r)$$

and $\quad W(D_s') - W(D_r') = (k+2)K_{r,s} - (k+1)(s-r).$

The two last equations permit us to conclude that

[2d] $\quad W(E_s) - W(E_r)$ and $W(D_s') - W(D_r')$ have the same sign for all $r < s$.

From (i) and (ii) we can establish for $j = 1, \ldots, k+1$ that

[2e] $\quad$ if E_j has exceedance r_j, then D_j' has exceedance $r_j + 1$.

Apply now the induction hypothesis to $E_1, \ldots, E_{k+1}$. We are assured that if $W(E_{j_1}) > W(E_{j_2}) \ldots > W(E_{j_{k+1}})$ where $(j_1, \ldots, j_{k+1})$ is a permutation of $(1, \ldots, k+1)$, then the exceedances of $E_{j_1}, \ldots, E_{j_{k+1}}$ are $k, k-1, \ldots, 0$. Statements [2d] and [2e] now yield $W(D'_{j_1}) > \ldots > W(D'_{j_{k+1}})$, and the exceedances of $D'_{j_1}, \ldots, D'_{j_{k+1}}$ are $k+1, \ldots, 1$. It is almost immediate – through induction, for example – that $W(D_0') < W(D_j')$ $(j = 1, \ldots, k+1)$ and D_0' has exceedance 0, where D_0' is the cumulative composition of C_0. An application of Lemmas 2C and 2A proves that $W(A_i) > W(A_j)$ implies that $r_i > r_j$, as asserted. □

3
LATTICE PATHS AND THE BALLOT THEOREM

A good illustration of the interplay between probability theory and combinatorial analysis during the last twenty-five years is provided by the celebrated *problème du scrutin* or ballot theorem and its generalizations. After its elegant solution by D. André (1887), the ballot theorem was placed in its proper perspective by Feller (1950), and many generalizations and applications followed. As a masterly exposition of these combinatorial methods with detailed references and vast applications is contained in Takács (1967), we give two proofs of the ballot problem restricting ourselves to the combinatorial and lattice path aspects with which we are concerned.

THEOREM 3A (D. André) *Let m, n be integers satisfying $1 \leq n < m$. The number of lattice paths $N(\mathscr{P})$ joining O to the point (m, n) and not touching the diagonal $x = y$ except at O is given by*

$$[3a] \quad N(\mathscr{P}) = \frac{m-n}{m+n}\binom{m+n}{n}.$$

In other words, given a ballot at the end of which candidates P, Q obtain m, n votes, respectively, the probability that P leads Q throughout the counting of votes is $(m-n)/(m+n)$.

□Let $L(m,n)$ be the set of all paths from $(0,0)$ to (m,n). Let us count paths from $(0,0)$ to (m,n), $m > n+t$, not touching the line $x = y+t$, where t is a non-zero integer. Denote by $L(m,n;t)$ the above set of paths, and by $R(m,n;t)$ the subset of paths which reach the line $x = y+t$. Let $|L(m,n;t)|$, $|R(m,n;t)|$, and $|L(m,n)|$ be the number of paths in these sets. The *reflection principle*, as illustrated in Figure 1 (where t is negative), enables us to evaluate $|R(m,n;t)|$ by showing that there is a 1:1 correspondence between $R(m,n;t)$ and $L(m-t,n+t)$, i.e.

$$[3b] \quad R(m,n;t) \Leftrightarrow L(m-t,n+t).$$

Clearly every path in $R(m,n;t)$ must reach the line $x = y+t$ before reaching (m,n); consider a path reaching this 'boundary' for the *first* time at $(j,j-t)$ $(j = 0,1,\ldots,n+t)$. Reflect the portion of this path from $(0,0)$ to $(j,j-t)$ about the line $x = y+t$, so that the point (u,v), with $0 \leq u \leq j, 0 \leq v \leq j-t$, lying on the

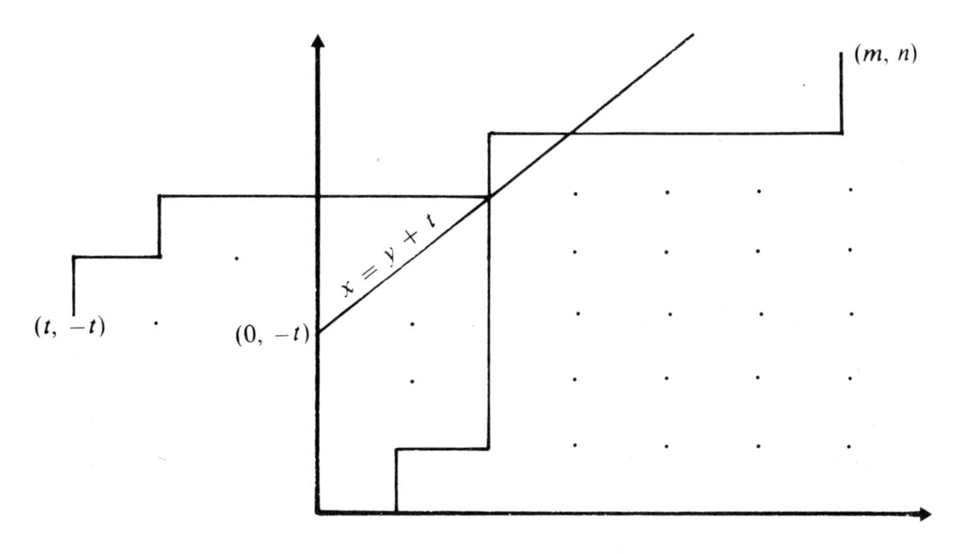

Figure 1

path becomes $(v+t, u-t)$. This reflected portion, together with the portion of the original path from $(j, j-t)$ to (m, n), constitutes a path from $(t, -t)$ to (m, n). We thus obtain [3b] and $|R(m, n; t)| = \binom{m+n}{n+t}$, since $L(m, n; t)$ and $R(m, n; t)$ partition $L(m, n)$,

$$|L(m, n; t)| = \binom{m+n}{n} - \binom{m+n}{m-t}.$$

To obtain Theorem 3A, we subtract from $\binom{m+n}{n} = |L(m, n)|$ all paths from $(0, 1)$ to (m, n) as well as all paths from $(1, 0)$ to (m, n) touching $x = y$. The number of subtracted paths is thus

$$L(m, n-1) + R(m-1, n, -1) = 2\binom{m+n-1}{n-1}.$$

We have [3a] immediately from

$$N(\mathscr{P}) = \binom{m+n}{n} - 2\binom{m+n-1}{n-1} = \frac{m-n}{m+n}\binom{m+n}{n}. \square$$

DEFINITION D Let $\mathscr{P}$, $\mathscr{Q}$ be two paths from $(0, 0)$ to (m, n). We say that $\mathscr{P}$ *dominates* $\mathscr{Q}$, written $\mathscr{P}\,\mathrm{D}\,\mathscr{Q}$, if no part of $\mathscr{Q}$ lies above $\mathscr{P}$. An illustration is

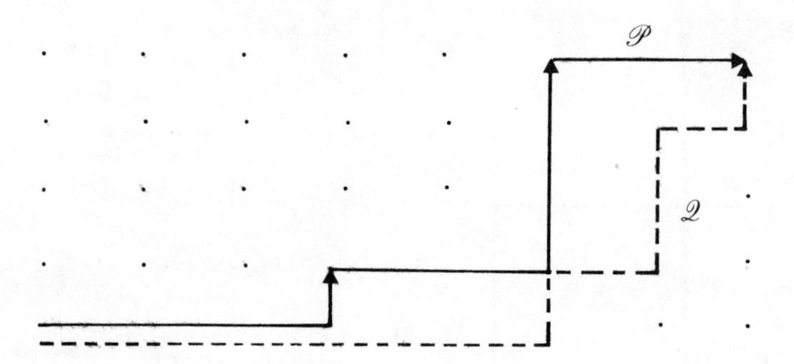

Figure 2

provided for $m = 7$, $n = 4$ in Figure 2. Clearly if $(a_1, \ldots, a_n)$, $(b_1, \ldots, b_n)$ are the vectors corresponding to $\mathscr{P}$, $\mathscr{Q}$ with

$$0 \le a_1 \le \ldots \le a_n \le m,$$

$$0 \le b_1 \le \ldots \le b_n \le m,$$

as in [1a], then $\mathscr{P}\,\mathrm{D}\,\mathscr{Q}$ if and only if

[3c] $\quad b_i \le a_i \qquad$ for $i = 1, \ldots, n$.

The number of vectors dominated by $\mathscr{P} = (a_1, \ldots, a_n)$ can be calculated by using the recursion formula stated as Lemma 3A below. Formula [3d] below follows from the inclusion-exclusion principle – see exercise 4 – but a simpler proof is provided in Chapter II.

LEMMA 3A *The number of paths dominated by the path $\mathscr{P}$ with vector $(a_1, \ldots, a_n)$ can be calculated as V_n using the recursion formula*

[3d] $$V_k = \sum_{j=1}^{k} (-1)^{j-1} \binom{a_{k-j+1}+1}{j} V_{k-j} = \binom{a_k+1}{1} V_{k-1}$$

$$- \binom{a_{k-1}+1}{2} V_{k-2} + \ldots + (-1)^{k-1} \binom{a_1+1}{k} V_0 \qquad \text{where } V_0 = 1.$$

Using the idea of domination, we prove the following theorem, which clearly reduces to Theorem 3A where $r = 1$.

THEOREM 3B (Ballot theorem) *Let m, n, r be positive integers satisfying $nr < m$. The number of paths joining O to the point (m, n) and not touching the line $x = ry$ except at O is given by*

[3e] $$N(\mathscr{P}) = \frac{m - rn}{m+n} \binom{m+n}{n}.$$

□Every path from $(0,0)$ to (m,n) not touching the boundary $x = ry$ except at O is dominated by

[3f] $[m-rn-1,\ m-r(n-1)-1, \dots, m-r-1]$.

Using the recursive formula [3d] on the vector [3f], or by direct calculation, we see that

$$V_1 = m-rn, \qquad V_2 = \frac{m-rn}{m+2-r(n-2)}\binom{m+2-r(n-2)}{2}.$$

Assuming inductively for values $i = 1, \dots, n-1$ that

[3g] $$V_i = \frac{m-rn}{m+i-r(n-i)}\binom{m+i-r(n-i)}{i},$$

we get, once again using [3d] with $k = n$,

[3h] $$\begin{aligned} V_n &= \sum_{j=1}^{n}(-1)^{j-1}\binom{m-rj}{j}\frac{m-rn}{m+n-j-rj}\binom{m+n-j-rj}{n-j} \\ &= \frac{m-rn}{n}\sum_{j=1}^{n}(-1)^{j-1}\binom{n}{j}\binom{m-rj-j+n-1}{n-1} \\ &= \frac{m-rn}{n}\left\{\sum_{j=0}^{n}(-1)^{j-1}\binom{n}{j}\binom{m-rj-j+n-1}{n-1}\right\}+\frac{m-rn}{n}\binom{m+n-1}{m-1}. \end{aligned}$$

The term within braces in the last line is zero, being the coefficient of $-x^m$ in $(1-x^{r+1})^n(1-x)^{-n}$, where, we recall, $m > rn$.

Thus

$$V_n = N(\mathscr{P}) = \frac{m-rn}{n}\binom{m+n-1}{n-1} = \frac{m-rn}{m+n}\binom{m+n}{n}. \square$$

We note that putting $n = i$ in [3g] yields the correct expression for $N(\mathscr{P})$ in [3e], so that we have essentially obtained the number of vectors dominated by $(a_1, \dots, a_n)$ where $a_1, \dots, a_n$ are in arithmetic progression. We state this result as Lemma 3B without proof, since it follows verbatim from Theorem 3B.

LEMMA 3B *The number of vectors dominated by* $(a_1, \dots, a_n)$ *where*

[3i] $a_i = a+(i-1)b \qquad (i = 1, \dots, n)$

and a, b *are non-negative integers is*

[3j] $$V_n(a,b) = \frac{a+1}{a+1+n(b+1)}\binom{a+1+n(b+1)}{n}.$$

Lemma 3B permits us to obtain the solution to any problem involving counting the set of vectors dominated by a vector $(a_1, \ldots, a_n)$ with the a_i's forming an arithmetic progression. As an illustration, if in Theorems 3A and 3B we change the words 'not touching' the boundary ($x = y$ or $x = ry$) to 'not crossing' (i.e. paths may touch the boundary but not cross it), we obtain effortlessly the number of such paths through Lemma 3B. Of course paths from $(0,0)$ to (m,n) not crossing $x = y+t$ are equivalent to those that do not touch $x = y+t+1$ if $t>0$ and that do not touch $x = y+t-1$ if $t<0$.

4
REPEATED REFLECTIONS AND APPLICATIONS

The reflection principle technique, used in Theorem 3A, can be extended to two straight-line boundaries, yielding results connected with the classical ruin problem in random walks which have been known for a long time. The same probabilities appear once again in certain (statistical) tests of the Kolmogorov-Smirnov type, and deep results originating with the work of N. Smirnov were simplified by B. V. Gnedenko around 1950 through an elegant geometric interpretation involving lattice paths. We content ourselves with a known example illustrating this interpretation, and provide at the end of the chapter a few selected references on this topic (see 4, 5, 7, and 10) for the reader who is interested in further applications to distribution-free tests or asymptotic expansions in probability theory.

By using the reflection principle of the last section repeatedly and a form of the inclusion-exclusion principle, we can evaluate the number of paths from the origin to (m,n) not touching two straight-line boundaries.

LEMMA 4A *Let $L(m,n;t,s)$ denote the set of paths from O to (m,n) not touching the lines $x = y+t$ and $x = y-s$ where $t,s>0$. Then*

$$[4a] \quad |L(m,n;t,s)| = \sum_k \left[\binom{m+n}{m-k(t+s)}_+ - \binom{m+n}{n+k(t+s)+t}_+ \right]$$

where

$$\binom{y}{z}_+ = \begin{cases} \binom{y}{z} & \text{if } y \geq z, \\ 0 & \text{if } z < 0 \text{ or } y < z, \end{cases}$$

and k takes all integer values, positive, negative, and zero.

□Let the boundaries $x = y+t$ and $x = y-s$ be denoted by T and S for simplicity. Let A_1 represent the set of paths reaching T at least once, regardless of what happens at any other step, and let A_2 denote the set of paths reaching T, S at least once in the order specified. Generally, A_i denotes the set of paths reaching

$T, S, T, \ldots$ (i times) at least once in the specified order. Let sets B_i be defined in the same way as A_i with S, T interchanged. A standard argument (see exercise 9) yields:

$$[4b] \quad |L(m,n;t,s)| = \binom{m+n}{n} + \sum_{i \geq 1} (-1)^i(|A_i| + |B_i|).$$

We evaluate $|A_i|$, $|B_i|$ by repeated reflection, using the notations of Theorem 3A and specifically relation [3b]. A typical evaluation is as follows: $A_3 \Leftrightarrow$ set of paths from $(t, -t)$ to (m, n) each of which reaches T after reaching S (by reflection about T) $\Leftrightarrow$ set of paths from $(-s-t, s+t)$ to (m, n) each of which reaches T (reflection about S) $\Leftrightarrow R(m+s+t,\ n-s-t;\ 2s+3t) \Leftrightarrow L(m-s-2t,\ n+s+2t)$. Thus $|A_3| = \binom{m+n}{m-s-2t}$ and generally

$$[4c] \quad |A_{2j}| = \binom{m+n}{m+j(s+t)}, \qquad |A_{2j+1}| = \binom{m+n}{m-j(s+t)-t}.$$

The expressions for $|B_{2j}|$ and $|B_{2j+1}|$ $(j = 0, 1, 2, \ldots,)$ – with $|A_0|$ and $|B_0|$ being $\binom{m+n}{n}$ – are obtained by interchanging m with n and s with t. Substitution of these values in [4b], after simplification, yields [4a].□

Let $x_1, \ldots, x_n$ and $y_1, \ldots, y_n$ be two independent random samples from the same continuous population $F(x)$. Let $F_n(x)$ and $G_n(x)$ denote the empirical distribution functions of the samples $x_1, \ldots, x_n$ and $y_1, \ldots, y_n$ respectively, i.e. $G_n(x)$ is the number of values $y_1, \ldots, y_n$ less than or equal to x divided by n, and similarly for $F_n(x)$. Combine the two samples as $X_1, \ldots, X_{2n}$ where the X's are in increasing order. Let X stand for a horizontal (vertical) step if X is an x (a y). Through use of this geometrical interpretation, the combined sample can represent all possible lattice paths from $(0, 0)$ to (n, n). These paths are equally likely since the independent samples are from the same population. Let

$$[4d] \quad D = D(n,n) = \sup_{-\infty < x < \infty} |F_n(x) - G_n(x)|.$$

Letting $P(S)$ stand for the probability of the event S, we see most easily from this geometrical interpretation that

$$\binom{2n}{n} \times P(D \leq d/n) = |L(n,n; d+1, d+1)|.$$

From [4a], we obtain immediately

$$[4e] \quad P(D \leq d/n) = \frac{1}{\binom{2n}{n}} \sum_k (-1)^k \binom{2n}{n-k(d+1)}.$$

If here $d = [(2n^{1/2}\lambda]$ and $n \to \infty$, we obtain (from Stirling's formula)

$$\lim_{n\to\infty} P[(n/2)^{1/2} D \leq \lambda] = \sum_{k=-\infty}^{\infty} (-1)^k e^{-2k^2\lambda^2},$$

a well-known special case of Smirnov's result.

We conclude with a word on notation. A path referred to in Definition A as $\mathscr{P}_0$ (script letter) may also be denoted as in Definition C (using the vectorial form) by A_0. No distinction need be made between these alternative notations as they are identical combinatorially. The former is, however, used by choice here when it is desirable to emphasize the polygonal line aspect as in Theorem 3A.

EXERCISES

1 Prove Lemma 2A by showing that $W(A_0), W(A_1), \ldots, W(A_n)$ form a complete set of residue classes modulo $n+1$. [*Hint*: Note, from [1f], that

$$\sum_{j=1}^{n} (a_j+i) = \sum_{j=1}^{n} a_j + ni$$

and the elements of A_i are precisely those of B_i, reduced possibly modulo $n+1$.]

2 Prove Lemma 2B by induction.

3 Complete the proof of Theorem 2A by proving the statement in the last but one sentence of Section 2: 'It is almost immediate that $W(D_0') < W(D_j') \ldots$'

4 Prove formula [3d]. *Hint*: Given $0 \leq a_1 \leq a_2 \leq \ldots \leq a_n$, let S_j $(j = 1, \ldots, n)$ be the set of vectors of integers satisfying (i) $0 \leq s_1 \leq \ldots \leq s_j$, (ii) $s_i \leq a_i$ $(i = 1, \ldots, j)$. Let $|S_k|$ or $N(S_k)$ denote the number of elements in a set. Then, by the inclusion-exclusion principle,

$$\begin{aligned} |S_n| = {} & N(s_n : s_n \leq a_n)|S_{n-1}| - N(s_{n-1}, s_n : s_n < s_{n-1} \leq a_{n-1})|S_{n-2}| \\ & + N(s_{n-2}, s_{n-1}, s_n : s_n < s_{n-1} < s_{n-2} \leq a_{n-2})|S_{n-3}| - \ldots \\ & + (-)^{n-1} N(s_1, \ldots, s_n : s_n < s_{n-1} < \ldots < s_1 \leq a_1). \end{aligned}$$

Notice that $N(s_{n-1}, s_n : s_n < s_{n-1} < a_{n-1}) = \sum_{s_{n-1}=1}^{a_{n-1}} \sum_{s_n=0}^{s_{n-1}-1} 1 = \binom{a_{n-1}+1}{2}$.

5 Verify Theorems 3A and 3B from Lemma 3B and obtain the corresponding results of these theorems when 'touching' is replaced by 'crossing.'

6 The identity $1 - \binom{n}{1} + \binom{n}{2} \ldots \pm (-1)^n \binom{n}{n} = 0$ may be restated as follows:

The vector $\left[1, -\binom{n}{1}, \binom{n}{2}, \ldots, (-1)^n\binom{n}{n}\right]$ is orthogonal to $[0^0, 1^0, \ldots, n^0]$. Show that it is orthogonal also to $[0^r, 1^r, \ldots, n^r]$ for $r = 1, \ldots, n-1$. More generally, for integral $n \geq 1$ prove that

$$\sum_{j=0}^{n} (-1)^j \binom{n}{j} (jx+n-1)_{(n-1)} = 0,$$

where

$$a_{(n)} = \begin{cases} a(a-1)\ldots(a-n+1) & \text{for integral } n > 0, \\ 1 & \text{for } n = 0. \end{cases}$$

7 *Compositions* An r-composition of an integer $N > 0$ is a vector $(t_1, \ldots, t_r)$ where the t_i's are positive integers satisfying $t_1 + \ldots + t_r = N$. Definition D of domination can be applied to the $\binom{N-1}{r-1}$ r-compositions of N as follows:

DEFINITION D $(t_1, \ldots, t_r)$ dominates $(t_1', \ldots, t_r')$ if and only if

$$\begin{aligned} t_1 &\geq t_1', \\ t_1 + t_2 &\geq t_1' + t_2', \\ &\vdots \\ t_1 + \ldots + t_{r-1} &\geq t_1' + \ldots + t_{r-1}'; \end{aligned}$$

of course

$$t_1 + \ldots + t_r = t_r' + \ldots + t_r' = N.$$

(An r-composition of M may dominate an r-composition of N, provided that $M \geq N$.)

Many results on compositions or distributing balls into ordered boxes, such as (a) and (b) below, can be obtained effortlessly from Lemmas 3A and 3B; but there are some surprises.

(a) Letting $T_j = t_1 + \ldots + t_j$ $(j = 1, \ldots, r)$ show that the number of r-compositions of N dominated by $(t_1, \ldots, t_r)$ is given by D_{r-1} in the recursive formula

$$D_k = \sum_{j=1}^{k} (-1)^{j-1} \binom{T_{k-j+1}+j-1}{j} D_{k-j} \qquad \text{with } D_0 = 1.$$

(b) Given that p, a, b, and $N-a-(p-1)b$ are all positive integers, show that the number of $(p+1)$-compositions of N dominated by the $(p+1)$-composition $(a, b, b, \ldots, b, N-a-(p-1)b)$ is

$$D_p(a, b) = \frac{a}{a+p(b-1)}\binom{a-1+pb}{p} = \frac{a}{a+pb}\binom{a+pb}{p}.$$

(c) Show that the r-compositions of N form a distributive lattice $(1 \leq r \leq N)$. [*Hint*: if $M_i = \max(t_1 + \ldots + t_i,\ t_1' + \ldots + t_i')$ $(i = 1, \ldots, r)$, then $(M_1, M_2 - M_1, \ldots, M_r - M_{r-1})$ is the least upper bound of $(t_1, \ldots, t_r)$, $(t_1', \ldots, t_r')$.]
(d) Let the complement of $(T_1, \ldots, T_{r-1})$ (see (a)) with respect to $(1, \ldots, N-1)$ be $(S_1, \ldots, S_{N-r})$. Show that the bijection

$$(T_1, \ldots, T_{r-1}, N) \leftrightarrow (S_1, \ldots, S_{N-r}, N)$$

defines an anti-isomorphism between the r-compositions and $(N-r+1)$-compositions of N.
(e) Let $a(N)$ and $b(N)$ be the set of all compositions of N with elements ≤ 2 and ≥ 2 respectively. Use the bijection in (d) to show that $|a(N)| = |b(N+2)|$, where $|S|$ is the number of elements in S.
(f) (L. Moser) Show that $|a(N)| = |a(N-1)| + |a(N-2)|$, so that both $|a(N)|$ and $|b(N)|$ for $N = 0, 1, 2, \ldots$ are related to the Fibonacci sequence $f(N)$ given by

$$f(0) = f(1) = 1,$$

$$f(N) = f(N-1) + f(N-2) \text{ for } N \geq 2.$$

8 Let $m = (n+1)k$ and consider the diagonal staircase from $(0,0)$ to (m,n) with horizontal steps of length k and vertical steps of length 1. The probability that a path from (0, 0) to (m, n) crosses the staircase is $\dfrac{nk-k-1}{nk+k+1}$. What is this probability when $m = nk + r$ and the first horizontal step is of length r and the rest of length k each?

9 Using the notations of Lemma 4A, partition the set of all paths $L(m,n)$ as follows: $\mathscr{A}_0$ = paths not reaching T or S, $\mathscr{A}_1$ = paths reaching T but not S, $\mathscr{A}_2$ = paths reaching T and S in that order once but not reaching T thereafter, etc.; the sets $\mathscr{B}_i$, $i = 1, 2, \ldots$, are defined by interchanging S and T.
(a) Show that $L(m,n) = \mathscr{A}_0 + \sum_{i=1}^{\infty} (\mathscr{A}_i + \mathscr{B}_i)$ and $A_{2i-1} + B_{2i-1} - A_{2i} - B_{2i} = \mathscr{A}_{2i-1} + \mathscr{B}_{2i-1} + \mathscr{A}_{2i} + \mathscr{B}_{2i}$ $(i \geq 1)$. Noting that $A_{2i-1} - A_{2i}$ and $B_{2i-1} - B_{2i}$ are disjoint, obtain [4b].
(b) Verify [4c] and the simplification in the last line of Lemma 4A.
(c) Prove that [4a] is the correct solution for $|L(m,n;t,s)|$ by showing that
(i) $|L(m,n;t,s)| = |L(m-1,n;t,s)| + |L(m,n-1;t,s)|$,
(ii) $|L(m,n;t,s)| = 0$ if $m = n+t$ or $n = m-s$.

10 (Poupard) Obtain a 1:1 correspondence between paths with exceedance 0 and paths with exceedance k where all paths from O to (n,n) are considered, thus re-proving the Chung-Feller Theorem.

11 (Kreweras) Let $W^+(A)$ be the area of path A above the diagonal if A is above the diagonal and 0 otherwise. Show that for the set S_n of all $\binom{2n}{n}$ paths in the $n \times n$ square

$$\sum_{A \in S_n} W^+(A_n) = 2^{2n-3}.$$

REFERENCES

1 ANDRÉ, D. 1887. 'Solution directe du problème résolu par M. Bertrand,' *Comp. Rend. Acad. Sci. Paris 105*, 436–7

2 BLACKMAN, J. 1956. 'An extension of the Kolmogorov distribution,' *Ann. Math. Statist. 27*, 513–20. Amended *Ann. Math. Statist. 29* (1958), 318–24

3 CHUNG, K. L., and W. FELLER. 1949. 'On fluctuations in coin tossing,' *Proc. Nat. Acad. Sci. U.S.A. 35*, 605–8

4 COMTET, L. 1970. *Analyse Combinatoire*. Presses Universitaire de France, Paris (English version: *Advanced Combinatorics*. D. Reidel, Dordrecht, 1974)

5 DRION, E. F. 1952. 'Some distribution-free tests for the difference between two empirical distribution functions,' *Ann. Math. Statist. 23*, 563–74

6 FELLER, W. 1968. *An Introduction to Probability Theory and Its Applications*. Vol. I, 3rd ed. (1st ed., 1950). John Wiley and Sons, New York

7 GNEDENKO, B. V., and V. S. KOROLJUK. 1951. 'On the maximum discrepancy between two empirical distribution functions,' *Dokl. Akad. Nauk S.S.S.R. 80*, 525–8 (in Russian)

8 HARDY, G. H., J. LITTLEWOOD, and G. POLYA. 1952. *Inequalities*. Cambridge University Press, London

9 POUPARD, C. 1967. 'Dénombrement de chemins minimaux à sauts imposés et de surdiagonalité donnée,' *Comp. Rend. Acad. Sci. Paris 264*, 167–9

10 SMIRNOV, N. 1939. 'On the estimation of the discrepancy between empirical curves of distribution for two independent samples,' *Bull. Math. Univ. Moscou*, Ser. A, *2*, No. 2, 3–14

11 TAKÁCS, L. 1967. *Combinatorial Methods in the Theory of Stochastic Processes*. John Wiley and Sons, New York

II

The dominance theorem and Smirnov test-statistics

L'homme n'est qu'un roseau, le plus faible de la nature,
mais c'est un roseau pensant.

Pascal, *Pensées*

Many interesting lattice path problems can be formulated in terms of enumerating lattice paths under certain restrictions, such as by avoiding certain boundaries. The simplest boundaries are, of course, straight lines and even here the introduction of the relation of domination (Definition A, Chapter I) permits us to obtain explicit results effortlessly. The most general two-boundary problem can be described, from this point of view, in terms of paths simultaneously dominating a vector $(b_1, \ldots, b_n)$ and dominated by a vector $(a_1, \ldots, a_n)$ with minimal restrictions on the a's and b's. We prove a general theorem for such paths by means of a simple geometric interpretation and study a few applications of domination in related combinatorial problems.

In Section 3, the distribution problem of the test-functions (or statistics) introduced by Smirnov for the 'two-sample problem' in 1939 is shown to be intimately related with dominance. After much work devoted by statisticians to studying and tabling the Smirnov distributions, G. P. Steck (1969) explicitly proved in the general case $m \neq n$ that the difficult distribution problem is related to dominance of ranks. Following Lehmann, Steck later (1974) gave a closed expression for the probability distribution of rank dominance when the two samples are not identically distributed but related by the condition $F = G^k$. In statistical terminology, the power against Lehmann alternatives of tests whose critical or acceptance regions consist of paths dominated by a given path is known; leaving such statistical aspects to the next chapter, we conclude with some combinatorial results motivating recent developments connected with the Smirnov tests.

1
A THEOREM ON DOMINATION AND ITS GEOMETRICAL INTERPRETATION

THEOREM 1A (Kreweras 1965) *Let* $0 \leq b_1 \leq \ldots \leq b_n$ *and* $0 \leq a_1 \leq \ldots \leq a_n$ *be two sets of integers satisfying* $b_i \leq a_i$ $(i = 1, \ldots, n)$. *Let* $s^{(j)} = (s_{1j}, \ldots, s_{nj})$, $j = 1, 2, \ldots,$ *be a set of vectors satisfying the inequalities*

$$[1a] \quad 0 \leq s_{1j} \leq \ldots \leq s_{nj}, \quad b_i \leq s_{ij} \leq s_{i,j+1} \leq a_i \qquad (j = 1, 2, \ldots; i = 1, \ldots, n).$$

If $|b, a; r)|$ *denotes the number of* $n \times r$ *matrices* $[s_{ij}]$ *satisfying* [1a], *then for* $r = 1, 2, \ldots$

$$[1b] \quad |(b, a; r)| = \text{determinant } c_{ij}^{(r)} = \|c_{ij}^{(r)}\|,$$

where

$$[1c] \quad c_{ij}^{(r)} = \binom{a_i - b_j + r}{r + j - i}_+ \quad \text{or} \quad c_{ij}^{(r)} = \binom{a_{n-j+1} - b_{n-i+1} + r}{r + j - i}_+$$

and

$$[1d] \quad \binom{y}{z}_+ = \begin{cases} \binom{y}{z} & \text{if } y \geq z, \\ 0 & \text{if } y < z \text{ or } z < 0. \end{cases}$$

Geometrically, if a particle moves in n-dimensions from the lattice point $b = (b_1, \ldots, b_n)$ *to* $a = (a_1, \ldots, a_n)$ *stopping at* r *intermediate points* $s^{(1)}, \ldots, s^{(r)}$ *satisfying* [1a], *we count in how many different ways this is possible.*

□It is natural to set $|(b, a; 0)| = 1$ so that [1b] is true for $r = 0$ with

$$c_{ij}^{(0)} = \binom{a_{n-j+1} - b_{n-i+1}}{j - i}_+.$$

Now

$$|(b, a; 1)| = \sum_{\substack{\text{over all} \\ \text{permissible } s^{(1)}}} |s^{(1)}, a; 0)|,$$

since the term inside the summation is 1. Thus

$$[1e] \quad |b, a; 1)| = \sum_{s_{n1} = y_n}^{a_n} \sum_{s_{n-1,1} = y_{n-1}}^{a_{n-1}} \cdots \sum_{s_{11} = b_1}^{a_1} \left\| \binom{s_{n-j+1,1} - b_{n-i+1}}{j - i}_+ \right\|$$

and

$$y_k = \max(b_k, s_{k-1,1}) \qquad (k = 2, \ldots, n).$$

Assume that $y_n = s_{n-1,1}$ and note that s_{n1} occurs only in the first column of the determinant. The sum over s_{n1} thus yields a determinant with first column

$$[1f] \quad \binom{a_n - b_{n-i+1} + 1}{1 - i + 1}_+ - \binom{s_{n-1,1} - b_{n-i+1}}{1 - i + 1}_+,$$

the remaining columns being unchanged. The negative terms in [1f] are exactly the same as the terms in column 2 of our determinant with sign changed, so these terms can be dropped out of column 1. Even if $y_n = b_n$, so that $b_n = \max(s_{n-1,1}, b_n)$, the negative term in row 1 as well as all other terms in row 1 (except in column 1) are zero. So, in both cases, the result of the first summation can be written as a determinant with $\binom{a_n - b_{n-i+1} + 1}{1 - i + 1}_+$ in column 1. A very similar argument regarding summation applies to $s_{n-1,1}, \ldots, s_{11}$, and the other columns, so that

$$[1g] \quad |(b, a; 1)| = \left\| \binom{a_{n-j+1} - b_{n-i+1} + 1}{1 + j - i}_+ \right\|,$$

and the theorem is true for $r = 1$ with this $c_{ij}^{(1)}$. (A direct proof of [1g] is asked in the exercises.)

We now assume the theorem is true for all integers up to $r-1$, so that

$$[h] \quad |(b, a; r-1)| = \left\| \binom{a_{n-j+1} - b_{n-i+1} + r - 1}{r - 1 + j - i}_+ \right\|.$$

Noting, most easily from the geometrical interpretation, that

$$[1i] \quad |(b, a; r)| = \sum_{s_{nr}=u_n}^{a_n} \cdots \sum_{s_{2r}=u_2}^{a_2} \sum_{s_{1r}=b_1}^{a_1} |s^{(r)}, a; r-1)|$$

with $u_k = \max(b_k, s_{k-1,r})$ $(k = 2, \ldots, n)$, we have the identity corresponding to [1e] with $r > 1$. We repeat essentially the argument following [1e]; for $t \geq 2$, s_{tr} occurs in column $n - t + 1$ only, and we note the identity

$$[1j] \quad \sum_{s_{tr}=u_t}^{a_t} \binom{s_{tr} - b_{n-i+1} + r - 1}{r + n - t - i}_+ = \binom{a_t - b_{n-i+1} + r}{1 + r + n - t - i}_+ - \binom{u_t - b_{n-i+1} + r - 1}{1 + r + n - t - i}_+.$$

Precisely as after [1f] and by the same simple properties of determinants, the negative term in [1j] is dropped if $s_{t-1,r} \geq b_t$; and if $b_t > s_{t-1,r}$ $(t \geq 2)$, this negative term in row $n - t + 1$ is zero, as well as all terms of row $n - t + 1$ in columns $n - t + 2, \ldots, n$. Indeed the whole upper right-hand matrix of $n - t + 1$ rows and $t - 1$ columns is zero. Thus, we conclude in both cases, by means of simple properties of determinants and by dropping obviously unimportant terms, that

$$|b, a; r)| = \left\| \binom{a_{n-j+1} - b_{n-i+1} + r}{r + j - i}_+ \right\|.$$

The induction is now complete. Finally,

$$\left\| \binom{a_i - b_j + r}{r + j - i}_+ \right\| = \left\| \binom{a_{n-j+1} - b_{n-i+1} + r}{r + j - i}_+ \right\|$$

so that the alternative form in [1c] is also valid. □

2
DISCUSSION OF THE DOMINANCE THEOREM AND SOME SPECIAL CASES

Enumeration problems of lattice paths correspond to the special case $r = 1$ of Theorem 1A. This special case was rediscovered independently by Steck (1969); the present inductive proof is a straightforward adaptation of a 'one-sided' proof by Narayana (1955). The general result for $r \geq 1$ is no more difficult to obtain than that for $r = 1$ and much simpler than solving each special enumeration problem anew. Apart from this economy, we obtain a new class of combinatorial identities from the general case. We also discuss a few easy variants of Theorem 1A which present no new difficulties.

2.1. *The case $b = 0$*

Let $r = 1$ and $b_1 = \dots = b_n = 0$. The determinant $\left\| \binom{a_{n-j+1}+1}{1+j-i}_+ \right\|$ can be written out in full as

$$[2a] \quad V_n = \begin{vmatrix} \binom{a_n+1}{1} & \binom{a_{n-1}+1}{2} & \cdots & \binom{a_2+1}{n-1} & \binom{a_1+1}{n} \\ 1 & \binom{a_{n-1}+1}{1} & \cdots & \binom{a_2+1}{n-2} & \binom{a_1+1}{n-1} \\ 0 & 1 & \cdots & \vdots & \vdots \\ \vdots & 0 & \cdots & \vdots & \vdots \\ 0 & 0 & & 1 & \binom{a_1+1}{1} \end{vmatrix}.$$

Expanding V_n by its first column gives

$$[2b] \quad V_n = \binom{a_n+1}{1} V_{n-1} - R_{n-1}$$

where R_{n-1} is V_n with column 1 and row 2 deleted. Repeated expansion of R_{n-1}, $R_{n-2}, \dots$ in terms of their first columns clearly yields [3d] of Chapter I, so that Lemma 3A, Chapter I, is a special case of Theorem 1A above.

2.2. *The duality principle with general r*

The simplest case of the duality principle occurs when $a_1 = \dots = a_n = m$ and $b_1 = \dots b_n = 0$, so that from our geometrical interpretation of Theorem 1A, $\det_{n \times n} \left| \binom{m+r}{r+j-i}_+ \right|$, with $i,j = 1, \dots, n$, counts the number of intermediate paths 'sandwiched' between b and a following [1a] above, i.e. in the rectangle shown in Figure 3. This number of intermediate 'sandwiched' paths is patently the same as

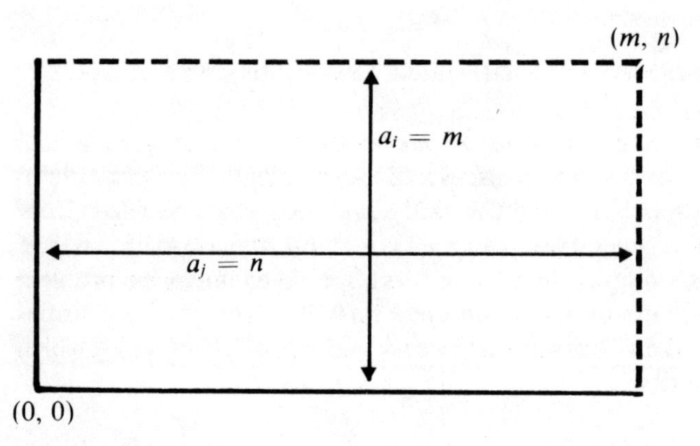

Figure 3

those between $a_1 = \dots = a_m = n$ and $b_1 = \dots = b_m = 0$, namely

$$\det_{m \times m} \left| \binom{n+r}{r+j-1}_+ \right| \qquad \text{with } i,j = 1, \dots, m.$$

Hence

$$[2c] \quad \det_{n \times n} \left| \binom{m+r}{r+j-i}_+ \right| = \det_{m \times m} \left| \binom{n+r}{r+j-i}_+ \right|$$

with general $m, n, r > 0$. Although the identity is trivial if $m = n$, $m \neq n$ gives new results.

We shall quote only one further identity corresponding to the repeated reflection lemma, Lemma 4A of Chapter I. It is clear that we can obtain a large class of determinantal identities of the same character as [2d] below (which reduces to (2c) for $s = t = \infty$) by duality.

Let $m+s > n > m-t$, where m, n, s, t, r are positive integers. Then

$$[2d] \quad \det_{m \times m} \left| \binom{n-(n-s+2-i)_+ - (j-t+1)_+ + r}{r+j-i}_+ \right|$$

$$= \det_{n \times n} \left| \binom{m-(n-s+2-i)_+ - (j-n+m-t+1)_+ + r}{r+j-1}_+ \right|$$

where

$$(a)_+ = \begin{cases} a & \text{if } a \geq 0, \\ 0 & \text{otherwise.} \end{cases}$$

2.3. *A modification of the dominance theorem*

It is possible to modify Theorem 1A to give explicit solutions to similar enumeration problems; indeed the geometrical interpretation and enumeration technique remain essentially the same as before. However, one fundamental difference between Theorem 2A below, which represents a slight modification, and Theorem 1A is that no duality is available for Theorem 2A.

THEOREM 2A *A particle starts from the origin in n-dimensional space and reaches the point* $(a_1, \ldots, a_n)$ *in r steps where* $1 \le r \le a_1 \ldots \le a_n$, *according to the following scheme. Let* $a_{i\alpha} \ge 1$ *be the increase in the ith coordinate at step* α. *Let the* $a_{i\alpha}$ *satisfy*

$$[2e] \quad \begin{aligned} a_{n1} \ge \ldots \ge a_{11} \ge 1, \\ (a_{n1}+a_{n2}) \ge \ldots \ge (a_{11}+a_{12}) \ge 2, \\ \vdots \\ (a_{n1}+\ldots+a_{n,r-1}) \ge \ldots \ge (a_{11}+\ldots+a_{1,r-1}) \ge r-1. \end{aligned}$$

Of course $a_n = (a_{n1}+\ldots+a_{nr}) \ge \ldots \ge (a_{11}+\ldots+a_{1r}) = a_1 \ge r$. *Then if* $(a_1, \ldots, a_n)_r$ *is the number of ways in which the particle can move in this way, we have*

$$[2f] \quad (a_1, \ldots, a_n)_r = \det_{n \times n} \left| \binom{a_{n-i+1}-1}{r-1+i-j}_+ \right| = \det_{n \times n} \left| \binom{a_j-1}{r-1+i-j}_+ \right|.$$

□The inductive proof starting with $(a_1, \ldots, a_n)_1 = 1$ is almost identical with that of Theorem 1A (perhaps slightly simpler) and is therefore omitted.□

The reader is invited to state Theorem 2A in terms of matrices $[s_{ij}]$ satisfying conditions similar to [1a] above, or to state Theorem 1A with $a_{i\alpha} \ge 0$. Despite the very similar geometrical interpretations, we note that the passage problem from $(b_1, \ldots, b_n)$ to $(a_1, \ldots, a_n)$ in Theorem 2A is not serious. Even assuming that $0 \le b_1 \le \ldots \le b_n$, we need the further condition

$$[2g] \quad 0 \le b_1 - a_1 \le b_2 - a_2 \le \ldots \le b_n - a_n$$

in order for the trivial modification of [2f] to hold true. Thus unlike Theorem 1A, where the point $b = (b_1, \ldots, b_n)$ plays an essential role, modifications of Theorem 2A without condition [2g] are false.

2.4. *Lattice paths with diagonal steps*

The case $n = 2$ of Theorem 2A yields once again proofs of the ballot theorem which can be adapted easily to lattice paths with diagonal steps in the plane. More general results in three dimensions follow from the same principles, although no systematic investigation in higher dimensions appears to have been carried out.

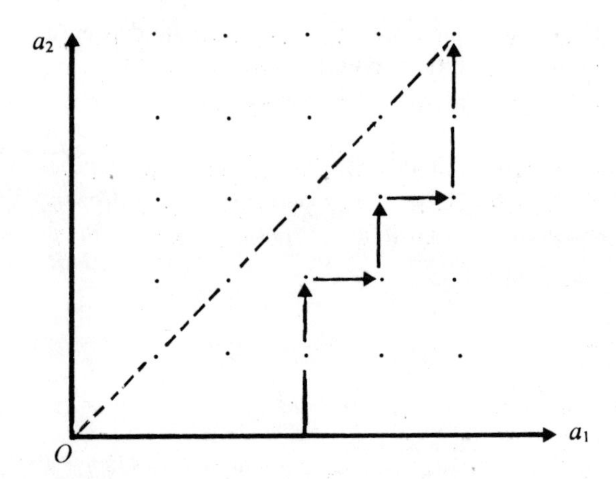

Figure 4

Letting $n = 2$ and $r = t$ in Theorem 2A we get

$$[2h] \quad (a_1, a_2)_t = \begin{vmatrix} \binom{a_1-1}{t-1} & \binom{a_2-1}{t-2} \\ \binom{a_1-1}{t} & \binom{a_2-1}{t-1} \end{vmatrix}$$

for $t \le a_1 \le a_2$. Setting $a_2 = a_1 = n$, we see (Figure 4 illustrates the fact that (3, 1, 1) dominates (2, 1, 2)) that each such path from O to (n, n) represents a domination of a t-composition of n by a t-composition of n and conversely. Hence

$$[2i] \quad (n, n)_t = \frac{\binom{n}{t}\binom{n}{t-1}}{n} \qquad (1 \le t \le n)$$

can also be interpreted as the number of weak election leads with t-turns or paths with t-turns not crossing the diagonal from (0, 0) to (n, n).

The modification of [2h], [2i] to r-dominations in the plane is also valid, generalizing the ballot theorem (Theorem 3B, Chapter I). In fact, analogous to [2e], given $m > rn$, let the t-composition of $m(a_{21}, \ldots, a_{2t})$ r-dominate the t-composition of $n(a_{11}, \ldots, a_{1t})$ if

$$[2j] \quad \sum_{i=1}^{k} a_{2i} \ge r \sum_{i=1}^{k} a_{1i} \qquad (k = 1, \ldots, t).$$

Then letting $(n, m)_t^{(r)}$ be the analogue of $(a_2, a_1)_t$ (with $r \ge 1$) in [2h], we have

$$[2k] \quad (n, m)_t^{(r)} = \binom{m-1}{t-1}\binom{n-1}{t-1} - r\binom{m-1}{t-2}\binom{n-1}{t}.$$

We indicate a typical generalization of the ballot problem using [2h], involving diagonal steps. Ignoring diagonal steps, let us define a turn as a horizontal step or steps followed by a vertical step or steps. (A sequence of vertical steps followed by horizontal steps does *not* constitute a turn, for convenience.) Then (see exercise 2) the number of paths with k turns and no diagonal steps to $(m-r, n-r)$ such that each path lies below $y = x$ except at O is

$$[2l]\quad \binom{n-r-1}{k-1}\sum_{i=2}^{m-n+1}\binom{m-r-i}{k-1}-\binom{n-r-1}{k}\sum_{i=2}^{m-n+1}\binom{m-r-i}{k-2}$$
$$=\binom{n-r-1}{k-1}\binom{m-r-1}{k}-\binom{n-r-1}{k}\binom{m-r-1}{k-1}.$$

LEMMA 2A *The number of paths from O to (m,n) lying below $y = x$ (except at O) with exactly $r\,(0\leq r\leq n)$ diagonal steps and $k\geq 1$ turns is*

$$[2m]\quad \binom{m+n-r-1}{r}\left\{\binom{n-r-1}{k-1}\binom{m-r-1}{k}-\binom{n-r-1}{k}\binom{m-r-1}{k-1}\right\}.$$

□Each path to $(m-r, n-r)$ with no diagonal steps passes through $m+n-2r+1$ lattice points, which, for convenience, we call 'cells.' Excluding the cell corresponding to the origin, we can place r diagonal steps in $m+n-2r$ cells in $\binom{m+n-r-1}{r}$ ways. Clearly such a placement of a diagonal step in a path below $x = y$ continues to keep the path below $x = y$. Thus from [2l] we obtain immediately [2m].□

Further details and three-dimensional problems are provided in exercises 3, 4, 5, 6.

3
DUALITY AND APPLICATION TO SMIRNOV STATISTICS

3.1. Various types of duality have arisen in studying lattice paths, rank-order statistics, etc., and we define them before introducing the Smirnov statistics. A lattice path to (m,n) can be summarized (see Figure 5 for $m = 5, n = 4$) by $n(m)$ and the vector $V = (v_1, \ldots, v_m)$ $(H = (h_1, \ldots, h_n))$ where v_i (h_j) are the vertical (horizontal) distances of the path from the x- (y-) axis. Letting

$$[3a]\quad r_i = v_i + i \quad (i = 1, \ldots, m), \qquad s_j = h_j + j \quad (j = 1, \ldots, n),$$

we obtain the *rank vectors* $R = (r_1, \ldots, r_m)$, $S = (s_1, \ldots, s_n)$. In statistics, R and S denote (in the combined ordered sample) the ranks of the two samples $x_1, \ldots, x_m$ and $y_1, \ldots, y_n$ from two populations with distribution functions $F(x)$ and $G(x)$,

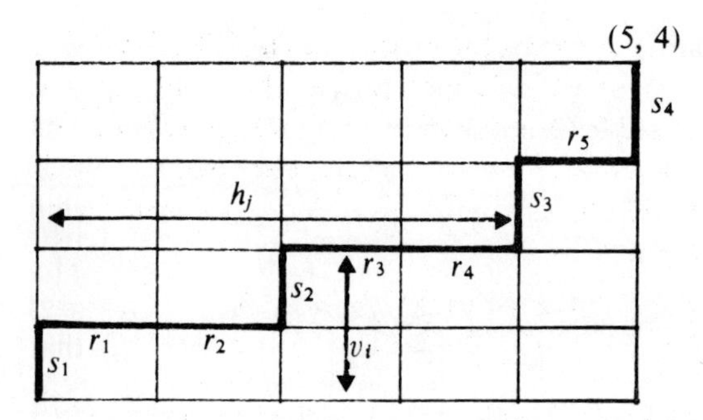

Figure 5

respectively. Evidently $R = \bar{S}$, where complementation is with respect to $1, \ldots, m+n$. Note that in [1a] of Chapter I we have defined a path to (m, n) using the vector $A = (m-h_n, \ldots, m-h_1)$. Given m, n, for the construction of certain tables, it makes for economy to assume, without loss of generality, that $m \geq n$ and to represent A *simply by its non-zero elements*. Since m, n are known, the path can be reconstructed with this information only. It is the need for tables which motivates our next definition, although it might appear most natural to use the obvious duality between H and V.

DEFINITION A (Path duality) The *dual* of a path $\mathscr{P}$ to (m, n) is the path $\mathscr{P}^*$ to (n, m) such that the vertical distances V of $\mathscr{P}$ become the distances A^* of $\mathscr{P}^*$. When $m = n$, $\mathscr{P}$ may coincide with $\mathscr{P}^*$ if it is symmetric about the diagonal $x+y = n$. A path is *self-dual*, when $m = n$, if $\mathscr{P} = \mathscr{P}^*$. Clearly the dual of a path $A = (a_1, \ldots, a_n)$ with $0 \leq a_1 \leq \ldots \leq a_n \leq m$ is the path A^* which has $m - a_n$ zeros, $a_n - a_{n-1}$ 1's, ..., a_1 n's. In Figure 5 $A = (0135)$, and for the dual path going to $(4, 5)$, $A^* = (11223)$. Also note that the set of all paths with ranks $(R_1, \ldots, R_m)$ satisfying $R_1 \leq r_1, \ldots, R_m \leq r_m$ is visibly identical with the set of paths dominated by the path with rank vector $R = (r_1, \ldots, r_m)$.

DEFINITION B (Lattice duality) The set of all paths with ranks dominating $(a_1, \ldots, a_m)$, i.e. $R_i \geq a_i$ $(i = 1, \ldots, m)$, is identical with the set of paths with ranks dominated by b_j, i.e. satisfying $S_j \leq b_j$ $(j = 1, \ldots, n)$, where $(a_1, \ldots, a_m)$, $(b_1, \ldots, b_n)$ are complementary sets with respect to $(1, \ldots, m+n)$.

For those dissatisfied with the visual proof implied by the statement made before the definition, very formal proofs have been obtained in exercise 7(d), Chapter I, covering even two-sided relations like $a_i \leq R_i \leq b_i$ and $c_j \leq S_j \leq d_j$. Of course Definition B applies to more complicated sets, such as to find the dual of $(R \geq a) \cup (R \geq b)$, and would apply equally well to sets of compositions, noting that the statement '$r_1, \ldots, r_m$ are ranks' is equivalent to '$(r_1, \ldots, r_m, m+n+1)$ is a $(m+1)$-composition of $m+n+1$.'

3.1. A further type of duality, which has been obtained by Lehmann by statistical methods, can be described as *probability duality under Lehmann alternatives.* As we do not prove technical statistical theorems in this book, we quote the following theorem as an example of a distribution on lattice paths with probability duality.

THEOREM 3A (Lehmann 1953) *Let* $R = (r_1, \ldots, r_m)$ *and* $S = (s_1, \ldots, s_n)$ *be the ranks of the two ordered samples of size m and n from the populations* $F(x)$ *and* $G(x)$, *respectively. Let* $k \geq 1$ *be an integer and let* $r_{m+1} = s_{n+1} = m+n+1$. *Then*

$$[3b] \quad P_{m,n}(S_1 = s_1, \ldots, S_n = s_n | G = F^k) = \frac{k^n}{\binom{m+n}{m}} \prod_{j=1}^{n} \frac{\Gamma(s_j + jk - j)}{\Gamma(s_{j+1} + jk - j)} \frac{\Gamma(s_{j+1})}{\Gamma(s_j)}$$

$$= P_{n,m}(S_1 = r_1, \ldots, S_m = r_m | G = F^{1/k})$$

$$= \frac{1}{k^m \binom{m+n}{n}} \prod_{i=1}^{m} \frac{\Gamma(r_i + i/k - i)}{\Gamma(r_{i+1} + i/k - i)} \frac{\Gamma(r_{i+1})}{\Gamma(r_i)}.$$

Of course $R = \bar{S}$ and $(R_1, \ldots, R_m)$, $(S_1, \ldots, S_n)$ refer to the ranks as random variables. For completeness, we conclude with a definition of symmetry and a simple lemma on self-dual paths.

DEFINITION C The *symmetric vector* A_n^S of any vector $A_n = (a_1, \ldots, a_n)$ to (m, n) is given by

$$A_n^S = (m - a_n, \ldots, m - a_1).$$

Thus A_n^S is the reflection of A_n about the diagonal and $(A_n^S)^* = (A_n^*)^S$.

LEMMA 3A *The number of self-dual paths in the* $n \times n$ *lattice is* 2^n. *Let* Δ_n *be the set of all such self-dual paths, and* s_n *be a subset of* $O_n = \{1, 3, \ldots, 2n-1\}$, *the set of first n odd integers. If* $W(A_n)$ *is the sum of all elements in* $A_n \in D_n$, *and* $\sum(s_n)$ *is the sum of all elements in* s_n, *there exists a bijection between subsets* s_n *of* O_n *and paths* A_n *of* Δ_n *such that*

$$W(A_n) = \textstyle\sum(s_n).$$

□The lemma may be proved by induction, the bijection being given explicitly by

$$[3c] \quad A_n = (a_1, \ldots, a_n) \leftrightarrow s_n = \{(2a_n - 1)_+, (2a_{n-1} - 3)_+, \ldots, (2a_1 - 2n - 1)_+\}$$

with $(a)_+ = a$ for $a \geq 0$ but 0 otherwise. The symmetry of the self-dual path about $x + y = n$ may be utilized, noting that $V_{a_n} = (0, \ldots, 0, 1, \ldots, 1, a_n)$ is a typical self-dual path, where there are $a_n - 1$ 1's in the n-vector V_{a_n}.□

3.2. The Smirnov statistics for independent samples from two populations $F(x)$, $G(x)$ are

$$D^+(m,n) = \sup_x (F_m(x) - G_n(x)),$$

$$[3d] \quad D^-(m,n) = \sup_x (G_n(x) - F_m(x)),$$

$$D(m,n) = \max(D^+(m,n), D^-(m,n)).$$

$F_m(x)$, $G_n(x)$ are the empirical distribution functions of the samples, defined as in Chapter I, where we discussed $D(n,n)$ when $F(x) = G(x)$. As explained in Chapter I, Gnedenko's lattice path interpretation applies to all three statistics in [3d], although we shall deal in this book only with $D^+(m,n)$. The work of many statisticians culminated in Steck's explicit relation between dominance and the Smirnov statistics $D^+(m,n)$, namely,

$$[3e] \quad \{mnD^+(m,n) \leq r\} = \{R_i \geq [i(m+n)-r]/m, 1 \leq i \leq m\}.$$

A simple illustration is given to clarify the combinatorial aspects; of course, $D^-(m,n)$ has a similar 'one-sided' relation with dominance, whereas $D(m,n)$ is a 'two-sided' statistic.

EXAMPLE Let $m = n = 4$. From [3e], $R_i \geq \max[i, (8i-r)/4]$, $1 \leq i \leq m$, so we obtain the following dominances for $r = 0, 1, \ldots, mn = 16$.

[3f]

r	R dominates ($\geq$)	A dominates ($\geq$)
0–3	2 4 6 8	1 2 3 4
4–7	1 3 5 7	1 2 3
8–11	1 2 4 6	1 2
12–15	1 2 3 5	1
≥ 16	1 2 3 4	0

The last column refers to the representation of A with non-zero elements (except for $0, 0, \ldots, 0 \equiv 0$!) which is most convenient for tables. It can be easily proved in general that the Smirnov distribution of $D^+(n,n)$ is equivalent to studying dominations of the self-dual sequence

$$[3g] \quad A_0 = 0 \ldots 0, \qquad A_i = 0 \ldots 012 \ldots i \quad (i = 1, \ldots, n),$$

where each A_i has n elements. Now whenever $W(A_{i+1}) > W(A_i) + 1$, dominance suggests we interpose 'intermediate' vectors between A_i and A_{i+1}. Letting $A^0 = A_i$, $A^n = A_{i+1}$, suppose such a set of intermediate 'sandwiched' vectors satisfy

$$[3h] \quad A^j \mathrm{D} A^{j-1}, \quad W(A^j) = W(A^{j-1}) + 1 \qquad (j = 1, \ldots, n).$$

The sandwiched vectors $A^1, \ldots, A^{n-1}$ form a *Young chain* (see, for example, reference 3) between A_i and A_{i+1} in combinatorial terminology. The number of Young chains between A_{n-1} and A_n is $n!$ (exercise 8) and such chains have long been studied in connection with the symmetric group (references 1, 9). From column A of [3f] and [3g], if we interpose these particular Young chains Y_1, Y_2, Y_3 between A_1 and A_2, A_2 and A_3, A_3 and A_4 respectively, then the chain $A_0, A_1, Y_1, A_2, Y_2, A_3, Y_3, A_4$ would constitute a 'dominance refinement' of the statistic $D^+(4,4)$.

Combinatorially, the above illustration generalizes immediately to all m, n, thus linking the theory of Young chains to dominance refinements of $D^+(m,n)$. Our discussion of the Smirnov test in Chapter III will clarify why we study dominance refinements of $D^+(m,n)$ and – perhaps for the reader who is unfamiliar with Smirnov statistics – even the term *dominance refinement* itself. Such a reader should work out in detail the cases (a) $m = 8$, $n = 4$, and (b) $m = 5$, $n = 6$, before attempting exercise 9, which may help clarify the combinatorial aspects.

EXERCISES

1 Prove relation [1g] directly.

2 Obtain the proof of equation [2l].

3 Consider the ballot problem with ties, where $m-r$ votes are for A, $n-r$ for B, and r votes for both A and B, i.e. the case of r diagonal steps where $0 \le r \le n < m$. The probability that A leads B throughout the counting is $(m-n)/(m+n-r)$. [*Hint*: Two proofs can be given using either the reflection principle of Chapter I or Lemma 2A.]

4 Consider paths in E^3 with cube diagonal steps, i.e. from (m,n,k) to $(m+1, n+1, k+1)$, as well as steps parallel to the axes. Using Lemma 2A and the classical ballot problem, show that the number of paths from $(0,0,0)$ to (m,n,k), $m>n$, with r cube diagonal steps such that each component of each path lies entirely to the $(m,0,0)$ side of the diagonal plane $y=x$ is

$$\frac{(m-n)(m+n+k-2r-1)!}{(m-r)!(n-r)!(k-r)!r!}.$$

5 (a) Show that the number of paths from $(0,0,0)$ to (m,n,k), $m \ge n$, such that no path has a component on the non-$(m,0,0)$ side of the diagonal plane $y=x$ and there are r cube diagonal steps is

$$(1) \quad \frac{(m-n+1)!(m+n+k-2r)!}{(m-r+1)!(n-r)!(k-r)!r!}.$$

(b) Show that the number of paths from $(0,0,0)$ to (n,n,n) with r cube diagonal steps such that, excepting of course the end points, each component of each path lies entirely to the $(n,0,0)$ side of the diagonal plane $y=x$ is

(2) $$\frac{(3n-2r-2)!}{(n-r)^2[(n-r-1)!]^3 r!}$$

(c) Setting $m = n = k$ in (1) and summing over r from 0 to n, and summing (2) over r from 0 to $n-1$, yields expressions for the appropriate numbers of paths from $(0,0,0)$ to (n,n,n). (Some elementary algebra is required to show that the expressions on pages 656 and 658 in Stocks (1967) simplify to the corresponding expressions given by (c).)

6 Generalize exercise 9(c) of Chapter I to diagonal steps.

7 In the classical ballot problem, A may hold a L-lead over B, L being an integer such that $1 \leq L \leq m-n$. (Of course the first $L+1$ votes must be for A.) Show that the probability that A holds the L-lead over B is

$$\frac{m!(m+n-L)!(m-L-n+1)}{(m+n)!(m-L+1)!}.$$

Show that A holds a 2-lead over B with probability one-half if $m = (2+\sqrt{5})n$ when n is large.

8 (a) Show that the number of Young chains between $0\,1\,2 \ldots n-1$ and $1\,2\,3 \ldots n$ is $n!$

†(b) (Kreweras) If ADB where $A = (a_1, \ldots, a_n)$ and $B = (b_1, \ldots, b_n)$ satisfy

$$b_i \geq a_i + 1 \qquad (i = 1, \ldots, n),$$

find the number of Young chains $Y(A, B)$ between A and B.

9 We denote by $Y^+(m,n)$ a dominance refinement of $D^+(m,n)$ including the trivial refinement $D^+(m,n)$ itself.

(a) If $m \neq n$, obtain from a given dominance refinement $Y^+(m,n)$ the dual refinement $Y^+(n,m)$ [*Hint*: Path duality. This result is useful in preparing Smirnov tables for statisticians.]

(b) For $m = n \geq 2$, show that non-trivial refinements $Y^+(n,n)$ of $D^+(n,n)$ always exist.

†(c) (Conjecture) True refinements of $D^+(m,n)$ exist if and only if $d = (m,n) > 1$.

†(d) (Unsolved) Determine for $m \neq n$ the number of possible refinements $Y^+(m,n)$ of $D^+(m,n)$.

10 A Young tableau of shape $(\alpha_1, \ldots, \alpha_r)$, $\alpha_1 \geq \alpha_2 \geq \ldots \geq \alpha_r$, $\sum_{i=1}^{r} \alpha_i = k$, is an arrangement of $1, 2, \ldots, k$ in r rows, with the ith row having α_i cells (in the

manner shown below for $r = 2$) such that the cell entries in rows and columns are in increasing order.

1	2	4	6	7	10	$\alpha_1 = 6$
3	5	8	9			$\alpha_2 = 4$

(a) Show that there is a 1:1 correspondence between Young tableaux of shape (m, n), $m \geq n$, and a path from $(0, 0)$ to (m, n) not crossing the line $x = y$.
(b) In a Young tableau, the hook length of cell (i, j) denoted by h_{ij} is given by

$$h_{ij} = 1 + \text{number of cells to the right of cell } (i, j) + \text{number of cells below cell } (i, j).$$

Frame, Robinson, and Thrall (reference 1) show that the number of Young tableaux of shape $(\alpha_1, \ldots, \alpha_r)$ where $\sum \alpha_i = k$ equals

$$\frac{k!}{\prod_i \prod_j h_{ij}}. \tag{3}$$

Derive from (3) the number of paths from $(0, 0)$ to (m, n) not crossing the line $x = y$.

11 Show that for $m = 2$, $n = 3$ the Lehmann distributions on r_1, r_2 with $k = 1/3$ and $k = 3$ are given by (see Theorem 3A)

(r_1, r_2)	$k = 1/3$	$k = 3$
(1, 2)	20	2268
(1, 3)	30	1134
(1, 4)	42	810
(1, 5)	56	648
(2, 3)	90	378
(2, 4)	126	270
(2, 5)	168	216
(3, 4)	252	180
(3, 5)	336	144
(4, 5)	560	112
	1680	6260

12 In the notation of Lemma 3A, let $B_{l,n}$ be the number of elements in the set $\{A_n \in \Delta_n;\ W(A_n) = l\}$. Show that $B_{l,n}$, i.e. the areas under self-dual paths, are

asymptotically normally distributed. [*Hint*: if $P_n(t) = \sum B_{l,n}t^l$, the generating function of $B_{l,n}$, then $P_n(t) = \prod_{k=0}^{n-1} (1+t^{2k+1})$.] Verify also that

(a) $B_{l,n} = B_{n^2-l,n}$,
(b) $B_{l,n+1} = B_{l,n} + B_{(n+1)^2-l,n}$,
(c) $B_{l,n} \neq 0$ if and only if $0 \leq l \leq n^2$, $l \neq 2$, n^2-2.

REFERENCES

1 Frame, J. S., G. de B. Robinson, and R. M. Thrall. 1954. 'The hook graphs of the symmetric groups,' *Can. J. Math. 6*, 316–23
2 Goodman, E., and T. V. Narayana. 1969. 'Lattice paths with diagonal steps,' *Can. Math. Bull. 12*, 847–55
3 Kreweras, G. 1965. 'Sur une classe de problèmes de dénombrement liés au treillis des partitions de entiers,' *Cahiers du Bur. Univ. de Rech. Opér. 6*, 5–105
4 Lehmann, E. L. 1953. 'The power of rank tests,' *Ann. Math. Statist. 24*, 23–43
5 Narayana, T. V. 1955. 'A combinatorial problem and its application to probability theory I,' *J. Indian Soc. Agric. Statist. 7*, 169–78
6 Steck, G. P. 1969. 'The Smirnov tests as rank tests,' *Ann. Math. Statist. 40*, 1449–66
7 – 1974. 'A new formula for $P(R_i \leq b_i, 1 \leq i \leq m \mid m, n, F = G^k)$.' *Ann. Probability 2*, 155–60
8 Stocks, D. R. 1967. 'Lattice paths in E^3 with diagonal steps,' *Can Math. Bull. 10*, 653–8
9 Young, A. 1927. 'On quantitative substitutional analysis,' *Proc. London Math. Soc. 28*, 255–92

III
Some applications of dominance to statistical problems*

Kennst du den Berg und seinen Wolkensteg?
Das Maultier sucht im Nebel seinen Weg,
In Höhlen wohnt der Drachen alte Brut,
Es stürzt der Fels und über ihn die Flut.
Kennst du ihn wohl?

Goethe, *Wilhelm Meisters Lehrjahre*

After a brief discussion of the role of dominance in combinatorial problems, two applications of dominance to statistics are considered in this chapter. The first is a dominance approach to the two-sample problem, in particular dominance tests of hypotheses for Lehmann alternatives. These alternatives have been considered in Chapter II in connection with probability duality. The second application is to enumeration problems of simple sampling plans connected with DeGroot's Theorem. As both of these applications involve relatively specialized statistical topics, we assume that the reader has sufficient background knowledge of statistics and has consulted references 2, 4, and 5 at the end of this chapter, which provide a good perspective for our applications.

1
THE ROLE OF DOMINANCE IN COMBINATORIAL PROBLEMS

The word *dominance* appears to have been introduced in combinatorial problems by Landau (1953), who studied conditions for a score structure in tournament problems. If $T = (t_1, \ldots, t_n)$ with $t_1 \le t_2 \le \ldots \le t_n$ represents the ordered scores of n players in a round-robin tournament, where each individual encounter ends in a win (1 point) for one player and a loss (0 points) for the other, then Landau proved that the necessary and sufficient condition that T represents a score vector is that T dominates $\left[0, \binom{2}{2}, \binom{3}{2}, \ldots, \binom{n}{2}\right]$, i.e.

* The reader should be familiar with references 2, 4, and 5 at the end of this chapter in order to understand the motivation behind our applications. See also Notes and Solutions, Chapter III.

$$
[1a] \quad
\begin{aligned}
t_1 &\geq 0, \\
t_1 + t_2 &\geq 1, \\
&\vdots \\
t_1 + \ldots + t_{n-1} &\geq \binom{n-1}{2}, \\
t_1 + \ldots + t_n &\geq \binom{n}{2}.
\end{aligned}
$$

Of course the last inequality here becomes an equality as obviously

$$W(T) = t_1 + \ldots + t_n = \binom{n}{2}.$$

A very thorough account of combinatorial aspects of tournaments is given by J. W. Moon (1968), to whose delightful book we refer the reader as well as to a more recently published proof of Landau's theorem by Brauer, Gentry, and Shah (1968). A more general definition of domination, essentially consistent with Definition D, Chapter I, was introduced by Narayana (1955). Although domination is strictly a special case of Hardy, Littlewood, and Polya's majorization, it appears appropriate to retain the term dominance to distinguish the subdomain – where integers and combinatorics are essentially involved – from the vast domain of problems where majorization is useful.

An excellent illustration of dominance in combinatorics is the elegant simultaneous treatment by G. Kreweras (1967) of the problems of Simon Newcomb and Young. A full treatment of such topics is beyond the scope of these notes, however, so we restrict our exposition of Kreweras's Theorem to a numerical example.

EXAMPLE Let $a = (223)$ dominate $b = (012)$ with $W(a) = 7$ and $W(b) = 3$. Let us suppose all *Young chains* $Y(a, b)$ between a and b are listed, and let us partition $Y(a, b)$ according to the number of switchbacks r (see below) in each chain. Then clearly $|Y(a, b)| = |Y_s(a, b)| + |Y_{s-1}(a, b)| + \ldots + |Y_0(a, b)|$. In the list below, $|Y(a, b)| = 8$ and $s = 2$, the maximum number of switchbacks possible, giving $8 = 2 + 5 + 1$.

The Set $Y(a, b)$

223

123	123	123	123	123	123	$\underline{222}$	$\underline{222}$
$\underline{122}$	023	023	$\underline{113}$	113	$\underline{122}$	122	122
022	$\underline{022}$	013	013	$\underline{112}$	$\underline{112}$	$\underline{112}$	022

012

If U, V, W are three consecutive increasing members in a Young chain, a *switchback* occurs at V if the term to increase in order to pass from V to W has strictly greater index than the term to increase in order to pass from U to V. (All switchbacks have been underlined in the above list.)

Now calculate the values $K_1, \ldots, K_s$ where

$$K_r(a,b) = \left\| \binom{a_i - b_j + r}{r + j - i}_+ \right\| = \text{number of `}r\text{-dominance chains,'}$$

as given by the dominance theorem. Clearly in our example $K_0 = 1, K_1 = 10, K_2 = 42$, and $|Y_0(a,b)| = 1$, $|Y_1(a,b)| = 5$, $|Y_2(a,b)| = 2$. Then Kreweras's Theorem enables us to calculate the number of Young chains with r switchbacks from the number of r-dominance chains given by the dominance theorem; precisely

$$[1b] \quad |Y_r(a,b)| = \sum_{k=0}^{r} (-1)^k \binom{W(a) - W(b) + 1}{k} K_{r-k} \qquad (r = 1, \ldots, s).$$

For a proof of this beautiful theorem and a detailed discussion of how it deeply generalizes many classical combinatorial results we refer to Kreweras's own papers, including those listed at the end of the book.

2
DOMINANCE TESTS FOR LEHMANN ALTERNATIVES

2.1. *Dominance tests for the two-sample problem*

There are several non-parametric tests for testing the null hypothesis $H_0: F = G$ under the assumption of continuity of F and G. We refer to Hájek (1969) for an elementary exposition of these tests. In this section we propose a new class of tests based on dominance and describe, purely for illustrative purposes, a test based on the dominance number $V_n(a_1, \ldots, a_n)$ of a path $A = (a_1, \ldots, a_n)$. (See [2a], Chapter II.) The example chosen may appear artificial, but best motivates – in our view – the later applications for both combinatorist and statistician.

In what follows, the sample space consists of all $\binom{m+n}{n}$ lattice paths from (0, 0) to (m, n).

DEFINITION A (Gnedenko-Feller region) Given any fixed path L, the set of paths of the form

$$[2a] \quad \{A : A \leq L\} \text{ or } \{A : A < L\} \text{ or } \{A : A \geq L\} \text{ or } \{A : A > L\}$$

is called a *Gnedenko-Feller* (GF) region.

DEFINITION B (Dominance test) Any test procedure which uses a GF region or unions of GF regions either as critical or as acceptance region is called a *dominance test* (D-test).

Clearly all D-tests are non-parametric tests, since all $\binom{m+n}{n}$ paths are equally likely under H_0; they are also rank tests (not necessarily linear in ranks) since path domination is equivalent to rank domination.

Suppose we wish to test $H_0:F = G$ against $H_1:F>G$ at level α, so that the number of paths in the critical region should not exceed $K = \left[\binom{m+n}{n}\alpha\right]$. An appropriate D-test, involving one GF region only, consists of taking paths A defined by $\{A:A \leq c\}$, where $V_n(c)$, the number of paths dominated by c, is $\leq K$. In Table 1 we present 5% critical paths ($5 \leq m, n \leq 10$) for the above procedure, i.e. $\alpha = 0.05$. For comparison Table 1 includes critical values and numbers of paths for the Wilcoxon rank-sum test or U-test. (Exercise 1 states that $U = \sum_{i=1}^{m} R_i - \frac{m(m+1)}{2}$ = area under a path.)

EXAMPLE (Hájek 1969, p. 70) Two independent random samples of $m = n = 10$ are drawn from two normal populations with $\mu_1 = 500$, $\sigma = 100$ and $\mu_2 = 580$, $\sigma = 100$ respectively. The data are given below (the values are rounded off to the nearest integers):

x's: 458, 620, 552, 327, 406, 733, 430, 498, 505, 558

y's: 746, 599, 690, 502, 425, 556, 491, 642, 622, 533

We shall illustrate both the U-test and D-test for testing $H_0:\mu_1 = \mu_2$ vs. $H_1:\mu_1 < \mu_2$. The combined ranking of x's and y's is as follows:

x	x	y	x	x	y	x	y	x	y
327	406	425	430	458	491	498	502	505	533

x	y	x	y	x	y	y	y	x	y
552	556	558	599	620	622	642	690	733	746

U-TEST: $U = 86 - \frac{1}{2}m(m+1) = 31$ Since the 5% point A_{10} (from Table 1) is 27, we accept H_0 at 5% level of significance. In fact it is easy to find that $p(U \leq 31) = 0.083$ and $p(U < 31) = 0.072$. Consequently the U-test accepts H_0 at 5% level of significance irrespective of randomization.

D-TEST From Table 1, the 5% critical vector is 122347788 (shown in Figure 6) with corresponding D_n value 9193. Since the observed 'Gnedenko' path, 111234568, is dominated by the critical vector (see Figure 6), the null hypothesis is rejected at the 5% level. The actual level attained is 9193/184756 = 0.0497.

TABLE 1

5% critical paths of a dominance test

(m, n)	5% of $\binom{m+n}{n}$	Area of U-test: A_n	No. of paths with area $\leq A_n$	Critical path for D-test: C	No. of paths dominated by C
(5, 5)	12	4	12	1122	12
(6, 5)	23	5	19	1223	23
(7, 5)	39	6	29	136	37
(8, 5)	64	8	60	11244	63
(9, 5)	100	9	83	1266	100
(10, 5)	150	11	149	11159	150
(6, 6)	46	7	43	22233	46
(7, 6)	85	8	63	112234	85
(8, 6)	150	10	122	11178	149
(9, 6)	250	12	220	24446	250
(10, 6)	400	14	374	114466	395
(7, 7)	171	11	167	124444	170
(8, 7)	321	13	302	23348	320
(9, 7)	572	15	519	1111578	569
(10, 7)	972	17	854	34688	965
(8, 8)	643	15	534	334466	642
(9, 8)	1215	18	1127	123788	1211
(10, 8)	2187	20	1819	1135688	2175
(9, 9)	2431	21	2283	13566666	2429
(10, 9)	4618	24	4375	12444778	4604
(10, 10)	9237	27	8241	122347788	9193

NOTE Initial zeros of paths are omitted for convenience; for example $1122 \equiv 01122$ since the number of elements in the path vector is exactly $n = 5$ (see column 5).

We remark that the critical paths given in Table 1 are not necessarily the best, nor do we claim any optimal properties for the D-test. Indeed statisticians insist on 'nested' sets of Gnedenko-Feller regions as in the Smirnov test illustrated in [3f], Chapter II, and this is the reason why our D_n-test of this section is artificial and for illustrative purposes only. However D-tests, as seen from Table 1, attain better (or more) levels than the Wilcoxon test since the partial ordering of paths by dominance number is finer than the partial ordering introduced by area under the path. This feature is shared by the dominance test to be introduced against Lehmann alternatives in Section 2.2 below; another common feature is that critical regions are obtained by random search (as in Table 1) on the Young lattice using a computer 'dominance package.'

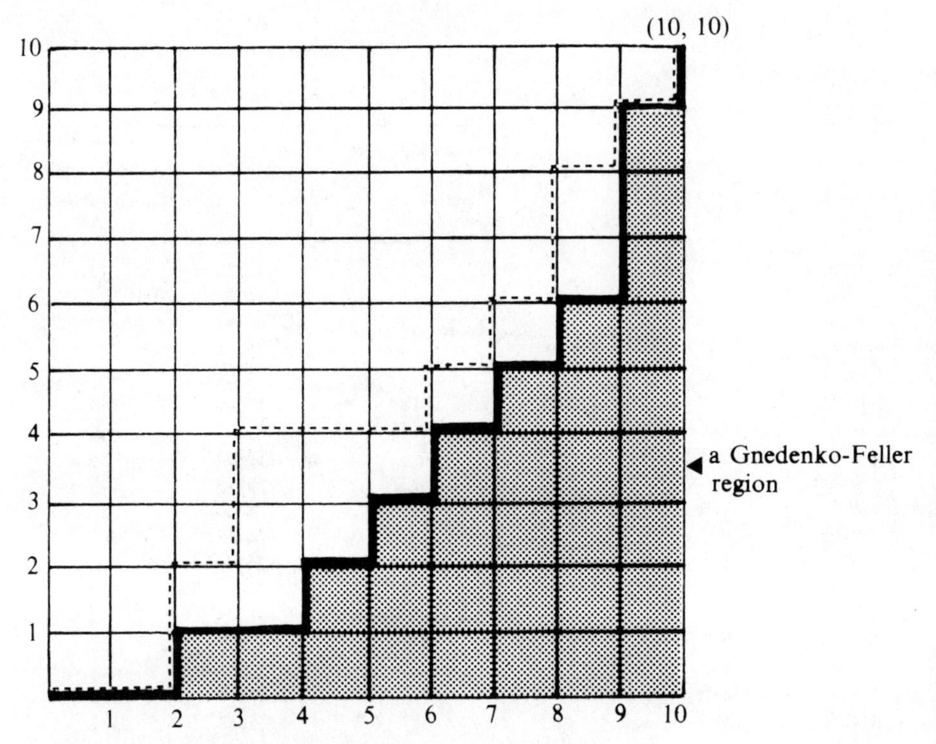

Figure 6. Illustration of the D-test. Note that the 5% critical path corresponding to the path 122347788 is shown by broken lines; the path represented by solid lines is the observed path 111234568 (initial zeros omitted).

2.2. *The UMP test against Lehmann alternatives*

Alternatives of the type $G(x) = F^k(x)$ with $k \geq 2$ integral are known to statisticians as Lehmann alternatives. The probabilities

$$[2b] \quad P_{m,n}(S_1 = s_1, \ldots, S_n = s_n) = \frac{k^n}{\binom{m+n}{n}} \prod_{j=1}^{n} \frac{\Gamma(s_j + kj - j)}{\Gamma(s_{j+1} + kj - j)} \frac{\Gamma(s_{j+1})}{\Gamma(s_j)}$$

with $r_{m+1} = s_{n+1} = m+n+1$ define a probability distribution on the $\binom{m+n}{n}$ lattice paths, attaching the probability [2b] to the path with ranks $s_1, \ldots, s_n$ (or horizontal ranks $r_1, \ldots, r_m$, where $r_1, \ldots, r_m = (s_1, \ldots, s_n)^c$ with respect to $1, \ldots, m+n$). Note that these probabilities are independent of F. As noted by Shorack (1967) and Steck (1974), these alternatives are appropriate in some biological and other problems, as well as being of exceptional combinatorial interest. For any fixed k, the uniformly most powerful (UMP) test for testing $F = G$ against $F = G^k$

can in principle be easily described: the UMP test of level α includes in the critical region those $K = \left[\alpha\binom{m+n}{n}\right]$ paths which have the highest probability.* However, without the concept of dominance, the listing of the K paths with highest probability is very tedious; indeed, the largest K for which the UMP test is explicitly known seems to be 93, in the case $m = n = 6$, $\alpha = 0.1$ (Lehmann 1953). The question arises whether some dominance approximation to the UMP test can be obtained for a larger range of values of m, n, say for $m, n \leq 30$.

The following lemma shows a relation between dominance and Lehmann alternatives, which is true for many other alternatives of interest as well.

LEMMA 2A *Suppose we are given two paths A, B with $A \mathrm{D} B$ and let $P_k(L) = P_{m,n}(R_1 = r_1, \ldots, R_m = r_m)$ be the probability of the path L as in* [2b]. *Then*

$$P_k(B) \geq P_k(A).$$

□See exercise 2.□

It immediately follows that every UMP test is a dominance test, since every UMP region must be a union of GF regions. (The reader is advised to work out a small numerical example, say $m = n = 3$, $k = 2$, completely; see also exercise 3 where some numerical information about the case $m = n = 6$ is provided.) It may thus be possible to approximate UMP tests with a dominance test involving only a small number of GF regions, say two or three, without losing much power. Indeed the power of such approximations can be calculated from a result due to Steck (1974), who essentially extends Lehmann's arguments to establish the following lemma.

LEMMA 2B *For $k > 0$, let $P_k(\leq L)$ be the sum of all the probabilities $P_k(L)$* (as in Lemma 2A and [2b]) *of paths dominated by L, i.e. the power of the GF region defined by $\{A : A \leq L\}$. Letting $l_1, \ldots, l_m$ be the horizontal ranks of L, then*

$$P_k(\leq L) = P_{m,n}(R_1 \leq l_1, \ldots, R_m \leq l_m \mid F = G^{1/k})$$

$$\text{[2c]} \qquad = \frac{n!}{(n+m/k)!} \left\| \binom{j}{j-i+1} \frac{\Gamma(l_i - i + 1 + j/k)}{\Gamma[l_i - (i-1)(1-1/k)]} \right\|_{m \times m}.$$

Lemma 2B immediately allows us to calculate the power of regions such as $\{A : A \leq L_1\} \cup \{A : A \leq L_2\}$, which we denote by $(\leq L_1 \cup \leq L_2)$. By inclusion-exclusion or a very elementary probability formula, we have the relation

$$\text{[2d]} \quad P_k(\leq L_1 \cup \leq L_2) = P_k(\leq L_1) + P_k(\leq L_2) - P_k(\leq \overline{L_1 \cup L_2})$$

where the meaning of $\leq \overline{L_1 \cup L_2}$ is obvious (see exercise 4).

* If, as occasionally happens, two or more paths have the same probability, the level can be attained by an arbitrary assignment; exact levels can be achieved by *randomization* (if necessary).

As a first approximation, we propose a class of tests $DL_k(2)$ for each k, involving only two GF regions. For $m, n \leq 8$, and small k, we have calculated the power of $DL_k(2)$ from [2d] and compared it with the power of the Wilcoxon test as tabulated by Shorack (1967). As these calculations reveal that the power of $DL_k(2)$ is uniformly higher than the power of the Wilcoxon or U-test in the range considered, we feel that $DL_k(2)$ has reasonably high power. Furthermore, starting from the approximation $DL_k(2)$, dominance considerations permitted us to obtain the UMP test explicitly in the range $m, n \leq 8$ for small k by a random search. We describe briefly $DL_k(2)$ and its role in obtaining the UMP test.

To define $DL_2(2)$ we start off with the two nested sequences of paths, where each path consists of n elements:

[2e]

$$0 \ldots 01,\ 0 \ldots 011, \ldots, 01 \ldots 1,\ 1 \ldots 1 \quad (U_1).$$

$$0 \ldots 02,\ 0 \ldots 012, \ldots, 01 \ldots 12,\ 01 \ldots 13 \quad (V_1).$$

For $i \geq 1$ define U_{i+1}, V_{i+1} recursively by $U_{i+1} = U_i + 1 \ldots 1$, $V_{i+1} = V_i + 1 \ldots 1$, and let $U = (U_1, U_2, \ldots) = (u_1, u_2, \ldots)$, where $u_1, u_2, \ldots$ are the individual paths in $U_1, U_2, \ldots$ in that order; similarly $V = (V_1, V_2, \ldots) = (v_1, v_2, \ldots)$. Then the GF regions of $DL_2(2)$ are defined by

$$0 \ldots 0,\ 0 \ldots 01,\ 0 \ldots 011,\ v_i \cup u_{i_j} \text{ for } i \geq 1,$$

where u_{i_j} is the element of U_i which satisfies

[2f] $\quad P_2(u_{i_j}) \geq P_2(v_i) \quad$ and $\quad P_2(u_{i_j+1}) < P_2(v_i).$

In Table 2 the dominance profile of $DL_2(2)$ is presented for $m = n = 6$; complete tables for $4 \leq m, n \leq 8$ are available with the author. For obtaining $DL_3(2)$, where now $k = 3$, the procedure is exactly similar, using equations [2e] and [2f], except that the probabilities P_2 in [2f] are replaced by P_3 since $k = 3$. The reader is advised to work out exercises 5 and 6 for a better understanding of the details in constructing Table 2. Using the fact that the GF region in $DL_2(2)$ ($DL_k(2)$) which belongs to sequence U provides a rough approximation to the lattice path L_α such that there are approximately $\left[\alpha\binom{m+n}{n}\right]$ paths with higher probability than $P_2(L_\alpha)$ ($P_k(L_\alpha)$), through a random search we obtain the UMP test explicitly. (In Table 2 the level 222233 or 222223 is a good approximation for obtaining the UMP test at $\alpha = 0.1$). The essential point about our random search is that Lemma 2B is crucial in our search for locating $P_k(L_\alpha)$: if $P_k(L) < P_k(L_\alpha)$ for some path L, then all paths A satisfying $A \mathrm{D} L$, i.e. $P_k(A) < P_k(L)$, are automatically dropped from our search procedure. Such details of this purely 'computer problem' of locating $P_k(L_\alpha)$ are omitted, and our solution to obtaining the UMP test at the 5%

TABLE 2

Dominance profile of $DL_2(2)$ for $m = n = 6$, for values of α up to 0.1

GF region	Level
0	1
1	2
11	3
$11 \cup 2$	4
$1111 \cup 12$	7
$11111 \cup 112$	9
$111111 \cup 1112$	11
$111111 \cup 11112$	12
$111122 \cup 11113$	23
$111122 \cup 111113$	24
$112222 \cup 111123$	36
$122222 \cup 111223$	42
$222222 \cup 112223$	46
$222222 \cup 122223$	48
$222233 \cup 122224$	84
$222333 \cup 222224$	95

level explicitly for $k = 2, 4 \leq m, n \leq 8$ is given in Table 3. To apply the UMP test of Table 3, we proceed as follows:

(i) Obtain the observed Gnedenko path in the usual way (see worked-out example in Section 2.1).
(ii) Calculate $S_{m,n} = s_1(s_2+1)\ldots(s_n+n-1)$ where $s_1, \ldots, s_n$ are the (vertical) ranks of the sample corresponding to $G = F^2$.
(iii) If the calculated $S_{m,n}$ is greater than or equals the corresponding value in the table (i.e. under m, n), we reject the hypothesis $F = G$; otherwise we accept this hypothesis.

We have included in Table 3 the critical path $S_{m,n}$ at 5% level for its curiosity if not duality value. Indeed, we note that it is relatively easy to construct $DL_k(2)$ or similar approximations to the UMP test involving only two or a small number of GF regions. The path duals of such regions constitute in turn dominance tests which might be useful against other types of alternatives. No work has been done to obtain the dominance profiles of other tests, such as the normal scores test, or to compare them with the known dominance profiles of $DL_k(2)$ or their duals. Of course the dominance profiles of the Wilcoxon and Smirnov tests can be easily constructed for small m, n as they are dominance tests.

TABLE 3

UMP test for Lehmann alternatives $G = F^2$ at 5% level

n \ m	4	5	6	7	8
4	2376	3465	5002	6720*	9072
	(11)	(1111)	(22)	(4)	(1123)
	2112	3360	4800	6720	9009
	(111)	(12)	(1112)	(23), (222)	(2222)
5	24960	38808	56700	80640*	114240*
	(1111)	(1112)	(1222)	(5)	(1115)
	23166	36288	54600	80640	114240
	(12)	(122)	(123)	(22222)	(1134), (1333)
6	308880	473088	734400*	1111968	1627920*
	(112)	(1222)	(1114)	(1115)	(145)
	289575	465696	734400	1105920	1627920
	(22)	(1113)	(1133)	(2224)	(1116)
7	4375800	7001280	11176704	17136000	25740288*
	(122)	(1111222)	(1112223)	(112234)	(11136)
	4112640	6967296	11162880	17093160	25740288
	(111112)	(223)	(1333)	(1122333)	(11155)
8	69672960	116121600*	185175900	293025600*	449862336
	(11111112)	(24), (233), (2223)	(125)	(11112225)	(111246)
	66512160	116121600	184864680	293025600	449433600
	(113)	(222222)	(222233)	(11112244)	(12333334)

SOURCE: Computed by Dr G. N. Pandya.
* These are 'equal' levels requiring arbitrary assignment.

EXPLANATION OF TABLE The UMP test for testing $F = G$ against $G = F^2$ is given by $s_1(s_2+1)\ldots(s_n+n-1) \geq C$, where C is chosen to satisfy the level condition. For a given (m, n), the first main entry in the table denotes the C value for which the level (number of paths in critical region) is $\left[(0.05)\binom{m+n}{n}\right]$ and the second main entry corresponds to the level $\left[(0.05)\binom{m+n}{n}\right]+1$. The numbers in parentheses denote path(s) $a_1 a_2 \ldots a_n$ (with initial zeros omitted) for which $s_1(s_2+1)\ldots(s_n+n-1)$ equals the corresponding C value. The two main entries facilitate randomization if the exact 5% level is to be achieved.

2.3. The Smirnov and dominance ratio tests

We have already discussed the Smirnov test as a dominance test in Chapter II. We confine our attention to $D^+(m,n)$, which is appropriate for testing $H_0{:}F=G$ against $H_1{:}F(x)>G(x)$ for at least one x. Lehmann alternatives are of this type and the rejection region for H_0 against H_1 is defined by

[2g] $\{mnD^+(m,n)/d>s_\alpha\}$

where s_α depends on the level and $d=(m,n)$. Since

$$mn\,D^+(m,n)=\sup_{1\le i\le m}\{(m+n)i-mR_i\},$$

we have

[2h] $\{mnD^+(m,n)/d\le s_\alpha\}=\{R_i\ge[i(m+n)-s_\alpha d]/m,\,1\le i\le m\}$

for the acceptance region. From [2h] and $R_i\ge i$, the acceptance regions of $D^+(m,n)$ are GF regions, consisting of only one GF region, and hence D^+ is a dominance test.

Note that for fixed m,n only a finite number of levels are attained by the $D^+(m,n)$ test and not every path can be a possible boundary path (of the acceptance region) of $D^+(m,n)$ because of the restriction [2h]. In Table 4 we have

TABLE 4

N^+-boundaries (of acceptance region) with number of paths in the critical region

$m=6, n=4$		$m=6, n=8$		$m=8, n=4$	
Boundaries	No. of paths	Boundaries	No. of paths	Boundaries	No. of paths
1	1	1	1	1	1
2	5	2	7	2	5
12	10	12	14	3*	15
13	19	13	34	13	21
23	31	23	64	14*	40
24*	47	123	91	24	55
		124	139	25*	86
		134	209		
		234	289		
		235*	384		
		1235	454		
Total paths	210	Total paths	3003	Total paths	495

* These paths are the interposed boundaries.

TABLE 4 (continued)

N^+-boundaries (of acceptance region) with number of paths in the critical region

Boundaries of $N^+(n, n)$	No. of $[n^+(n, n)]$ for $n = 3, \ldots, 10$							
	3	4	5	6	7	8	9	10
1	1	1	1	1	1	1	1	1
2*	4	5	6	7	8	9	10	11
12	6	8	10	12	14	16	18	20
13*	11	17	24	32	41	51	62	74
23*	13	23	36	52	71	93	118	146
123	15	28	45	66	91	120	153	190
124*				114	166	230	307	398
134*				156	246	365	517	706
234*				192	316	485	706	986
1234				220	364	560	816	1140
1235*						835	1245	1777
1245*						1135	1795	2701
1345*						1415	2335	3646
2345*						1655	2785	4416
12345						1820	3060	4845
12346*							4061	6483
12356*							5161	8628
12456*							6211	10828
13456*							7171	12853
23456*							7996	14503
123456							8568	15504
123457*								19144
123467*								23148
123567*								26998
124567*								30598
134567								33898
234567*								36758
1234567								38760
Total paths	20	70	252	924	3432	12870	48620	184756

* These paths are the interposed boundaries.

inserted paths between Smirnov boundaries for all cases with $4 \leq m, n \leq 8$, and m, n non-coprime. From the theory of Young chains such refinements are not generally unique and the choice of refinement or interposed boundaries will depend on the alternative. The boundaries interposed are not any arbitrary boundaries but those we feel are suitable against Lehmann alternatives. As seen from Table 5, at least in the case $3 \leq m = n \leq 10$ the randomized power of the refined Smirnov test, called $N^+(m, n)$, is uniformly greater in our range at all three

TABLE 5

Randomized power of Smirnov and refined Smirnov tests against Lehmann alternatives $G = F^k$ for $k = 2, 3$

Level of significance	$k=2$		$k=3$		$k=2$		$k=3$	
	D^+	N^+	D^+	N^+	D^+	N^+	D^+	N^+
	$m=3, n=3$				$m=4, n=4$			
0.01	—	—	—	—	—	—	—	—
0.05	—	—	—	—	.15834	.17708	.26029	.29969
0.10	.23382	.2500	.3409	.36817	.28664	.29663	.43779	.4588
	$m=5, n=5$				$m=6, n=6$			
0.01	.05262	.05758	.11103	.12402	.06763	.07222	.15353	.16833
0.05	.19409	.20250	.34082	.35604	.21404	.23620	.37901	.42681
0.10	.30213	.34578	.46642	.54501	.33996	.3672	.53739	.57943
	$m=7, n=7$				$m=8, n=8$			
0.01	.07397	.08693	.17261	.20965	.09191	.09381	.22650	.23126
0.05	.23359	.27002	.42219	.49382	.26759	.27946	.49170	.51196
0.10	.38174	.38952	.60548	.61869	.39471	.43255	.62499	.70166
	$m=9, n=9$				$m=10, n=10$			
0.01	.09776	.11275	.24141	.28464	.10970	.12848	.27796	.32975
0.05	.28622	.30805	.52405	.56455	.30277	.34057	.55592	.61982
0.10	.42054	.46543	.66701	.72818	.45478	.48454	.71714	.75444

levels of α considered. When m, n are coprime, as conjectured in Chapter II, exercise 9, no true refinement of $D^+(m, n)$ exists in the cases falling in our range, i.e. $4 \leq m, n \leq 8$. The problem of deciding which refinement to use for particular alternatives is open, and no claim as to the optimality of N^+ is made among all possible refinements.

For the combinatorist (or statistician) who considers randomization reprehensible in practical problems, we propose a test based on the dominance ratio (*DR*-test)

$$[2i] \quad DR(A) = DR(a_1, \ldots, a_n) = \frac{V(a_1, \ldots, a_n)}{V(m-a_n, \ldots, m-a_1)}.$$

TABLE 6

5% critical regions for the Wilcoxon and dominance ratio tests

$n = 3$, $m = 4$	$DR = W \leq 0$
$n = 3$, $m = 5$	$DR = W \leq 1$
$n = 3$, $m = 6$	$DR = W \leq 2$
$n = 3$, $m = 7$	$W \leq 2$*
	$DR = W + \{3, 12\}$
$n = 3$, $m = 8$	$W \leq 3$*
	$DR = W + \{4\}$
$n = 3$, $m = 9$	$W \leq 4$
	$DR = W - \{112\} + \{5\}$
$n = 4$, $m = 4$	$DR = W \leq 1$*
$n = 4$, $m = 5$	$W \leq 2$*
	$DR = W + \{3, 111\}$
$n = 4$, $m = 6$	$W \leq 3$*
	$DR = W + \{4, 22, 13\}$
$n = 4$, $m = 7$	$W \leq 4$*
	$DR = W + \{5, 23, 14, 122\}$
$n = 4$, $m = 8$	$W \leq 5$*
	$DR = W + \{6, 33, 15, 222, 24, 7, 114\}$
$n = 5$, $m = 5$	$DR = W \leq 5$
$n = 5$, $m = 6$	$W \leq 5$*
	$DR = W + \{33, 222, 24, 1122\}$
$n = 5$, $m = 7$	$W \leq 6$*
	$DR = W + \{34, 1222, 25, 133, 223, 16, 7, 115, 44, 1114\}$
$n = 6$, $m = 6$	$W \leq 7$*
	$DR = W - \{16, 111112,\} + \{44, 35, 233, 2222, 11222\}$

* Denotes cases where W does not attain the correct level.

Since a low dominance ratio would favour the alternative hypothesis $F(x) > G(x)$, for level α our critical region consists of those $K = \left[\alpha\binom{m+n}{n}\right]$ paths with least dominance ratio. The problem of explicitly determining the critical region for the *DR*-test is treated in a way that is rather similar to our discussion of the UMP test against Lehmann alternatives, since it is easily seen that the *DR*-test is also a dominance test. Table 6 presents the *DR*-test with the corresponding *U*-test or Wilcoxon test for comparison. Since the *DR*-test essentially achieves all levels, it presents an attractive alternative to the Wilcoxon test for the statistician who is opposed to randomization. Unlike the other tests proposed in this section, the *DR*-test is a competitor to the Wilcoxon test against *any* alternative.

2.4. *Characterization of simple sampling plans*

In this section we characterize simple sampling plans of size n by means of deformations which are closely related to Young chains. Unfortunately, as stated at the beginning of the chapter, the motivation for studying these statistical problems is fairly technical, and we assume the reader is thoroughly familiar with binomial sampling plans as first introduced by Girshick, Mosteller, and Savage (1946). Indeed, the combinatorist unaware of their statistical properties as discussed by DeGroot (1959) should skip this section entirely, except perhaps for the examples. For the statistician interested in obtaining a better perspective of these problems, we recommend Chapter 12 of Kagan, Linnik, and Rao's recent book (1973).

The definitions and notations we use follow DeGroot, with the following obvious extension: A boundary point (b.p.) γ_0 of a bounded sampling plan (s.p.) S is essential if the s.p. S_{γ_0}, where $S_{\gamma_0}(\gamma) = S(\gamma)$ for $\gamma = \gamma_0$, and $S_{\gamma_0}(\gamma_0) = 1$ (i.e., γ_0 becomes a continuation point (c.p.)), is not bounded. A b.p. is non-essential otherwise. It may be noted that the removal of any essential b.p. destroys closure in the sense of Lehmann and Stein (1950). A boundary is essential if all its b.ps. are essential.

Let $C_k = \{(x, y): x + y = k\}$, and let C be the set of all c.ps. and B be the set of all b.ps. in a s.p. S. Also, let $\gamma_0 = (x_0, y_0)$. Some important results describing the properties of a s.p. of size n, which follow as consequences of various definitions in DeGroot (1959) and the definition of an essential b.p., are given below.

REMARK 1 If γ_0 is a non-essential b.p. of a s.p. S of size n, the boundary of S contains more than $n+1$ points.

REMARK 2 For each k and for each pair γ_1, γ_2 of b.ps. of a simple s.p: S of size n on C_k, there are no inaccessible points (i.ps.) between γ_1 and γ_2. We note that the remark is true if either of the b.ps. γ_1, γ_2 is replaced by a c.p.

REMARK 3 If S is a simple s.p. of size n, then $C_n \cap B$ must form an interval containing at least two b.ps.

REMARK 4 If γ_0 is a b.p. of a s.p. S of size n, and if $\gamma_1 = (x_0, y_0 + 1)$ and $\gamma_2 = (x_0 + 1, y_0)$ are accessible, then γ_0 is a non-essential b.p.

REMARK 5 If γ_0 is a non-essential b.p. of a s.p. S of size n, then $C_{x_0+y_0} \cap C$ is not empty.

We now state a simple but fundamental lemma on simple s.ps.

LEMMA 2C *If S is a simple s.p. of size n, then all its b.ps. are essential.*

□Let $k > 1$ be the smallest integer such that C_k contains non-essential b.ps. of S. C_k contains at least one c.p. (Remark 5). Also $C_k \cap C$ is an interval since S is simple. Consider the lower end of $C_k \cap C$. At this end lies a b.p. which is either essential or non-essential (Remark 2). Let the b.p. be $P = (x, y)$ and let it be essential (see the figure below).

$$\begin{array}{cccc} & \cdot & \cdot\ P_u & \cdot \\ (x-1, y+1) & & & \\ & \cdot & \cdot\ P(x,y) & \cdot\ P_\gamma \\ & \cdot & \cdot & \cdot\ Q \end{array}$$

If P_u is a c.p., then P_γ must be an i.p. (Remark 4) and Q is either a b.p. or an i.p. If Q is an i.p. then all points below Q on C_k are i.ps. (Remark 2). On the other hand, if Q is a b.p., then it is essential. Consider the next lower point on C_k and repeat the argument until an i.p. is reached. The interval of b.ps. in this case is composed of essential b.ps. The argument holds good if P_u is a b.p. (P_u cannot be an i.p.).

Next assume $P = (x, y)$ to be non-essential. Then all points on C_k below P are either b.ps. or i.ps. Thus all points $(x+i, y), i = 1, 2, \ldots,$ are i.ps. In this case the s.p. S would be unbounded if the b.p. P is removed and hence P must be essential.

A similar argument shows that there cannot be any non-essential b.p. above the interval $C_k \cap C$ on C_k.□

We introduce the concept of deformation, which forms the basis of our characterization of simple s.ps. Let S be a bounded s.p. with essential boundary and γ_0 be a b.p. of S. Consider a new s.p. S' as follows:

$$S'(\gamma) = S(\gamma) \quad \text{for all } \gamma \neq \gamma_0, (x_0+1, y_0), (x_0, y_0+1),$$

and $S'(\gamma_0) = 1$. So far as (x_0+1, y_0) and (x_0, y_0+1) are concerned, only three cases can arise (Remark 4), viz.:

(i) both are inaccessible;
(ii) one is inaccessible, the other a b.p.;
(iii) one is inaccessible, the other a c.p.

In case (i), either $S'(x_0+1, y_0) = 0$ or $S'(x_0, y_0+1) = 0$, but not both; in case (ii), if γ_1 is the i.p. and γ_2 the b.p., then $S'(\gamma_1) = S'(\gamma_2) = 0$; in case (iii), if γ_1 is the i.p. and γ_2 the c.p., then $S'(\gamma_1) = 0, S'(\gamma_2) = 1$. In other words, S' is constructed by shifting the b.p γ_0 of S to one of the nearest i.ps. We then say that S' is a deformation of S at the b.p. γ_0. A deformation of S is possible at every b.p., and S' is an *admissible* deformation of S if S' is bounded and its boundary is essential.

Given any simple s.p. of size n, it is possible to get the single s.p. of size n (i.e., one whose b.ps. have index n), through a suitable choice of a series of admissible deformations.

THEOREM 2A *If S is a simple s.p. of size n, then there exists a sequence of s.ps. $S = S_0, S_1, S_2, \ldots, S_k$ such that S_{i+1} is an admissible deformation of S_i ($i = 0, \ldots, k-1$), S_k has exactly the points of index n as b.ps., and each S_i has $n+1$ b.ps.*

□Let B_0 be the boundary of S_0. Let k_0 be the smallest integer such that $C_{k_0} \cap B_0$ is not empty. (Assume $k_0 \neq n$.) The c.ps. on C_{k_0} form an interval A_0 which is not empty. If there is a contiguous b.p. $\gamma_0 = (x_0, y_0)$ on C_{k_0} above A_0, then (x_0+1, y_0) is an accessible point. Hence the deformation S_1 of S_0 at γ_0 has $S_1(\gamma) = S(\gamma)$ except for $\gamma = \gamma_0$, and $\gamma = (x_0, y_0+1)$. At the points γ_0 and (x_0, y_0+1), S_1 differs from S_0.

Above γ_0 on C_{k_0}, there are no i.ps. of S, because there are no b.ps. on lines C_k, $k < k_0$. Thus above y_0 on C_{k_0} are only b.ps. Since the only change in S_1 from S_0 is to shift the b.p. γ_0 to (x_0, y_0+1), making γ_0 a c.p., it is easily seen that S_1 is bounded if S_0 is bounded. Thus we have proved that S_1 is admissible.

In S_1, the interval of c.ps. on C_{k_0} contains one more point (viz. γ_0) than S_0 and the other c.ps. are those of S_0. A repetition of this process shifts the b.ps. one by one (whether from above or below A_0 on C_{k_0}) to the line C_{k_0+1}. If C_{k_0+1} has no c.ps., we have finished. Otherwise continue the process. Induction proves that we finally reach the region S_k where the b.ps. are exactly the points of index n. Clearly S_k and S_0 (in fact, every S_i) contain the same number of b.ps., namely $n+1$.□

We next prove the converse of Theorem 2A.

THEOREM 2B *If S is a non-simple s.p. of size n, then S contains more than n+1 b.ps.*

□Suppose S is of size n, non-simple, and has non-essential b.ps. Then Remark 1 proves the theorem. We restrict ourselves therefore to the case where S has only essential b.ps.

Let k_0 be the smallest integer such that C_{k_0} intersects the c.ps. of S in a configuration which is not an interval. Since the boundary of S is essential, the following configuration occurs in C_{k_0}: $\gamma_0 = (x_0, y_0)$ is a c.p.; $\gamma_i = (x_0+i, y_0-i) = (x_i, y_i)$, $i = 1, \ldots, k$, are b.ps.; and $\gamma_{k+1} = (x_0+k+1, y_0-k-1)$ is a c.p. where $k \geq 2$. Clearly $(x_1+ \; , y_1)$ is an i.p. and (x_1, y_1+1) is accessible. A deformation S_1 at γ_1 is of the form $S_1(\gamma) = S(\gamma)$ except for $\gamma = \gamma_1, \gamma = (x_1+1, y_1)$; at these points

$$S_1(\gamma_1) = 1, \qquad S_1(x_1+1, y_1) = 0.$$

The deformation S_1 is bounded because possible candidates for paths of arbitrary length must pass through γ_1, and any path through γ_1 either coincides with an S-path from (x_1, y_1+1) onwards or stops at (x_1+1, y_1).

In the same manner S_i can be constructed from S_{i-1}, $i = 2, \ldots, k-1$. In S_{k-1}, $\gamma_0, \gamma_1, \ldots, \gamma_{k-1}$ are c.ps., γ_k is a b.p., and γ_{k+1} is a c.p. Clearly γ_k is a non-essential b.p. of S_{k-1} which is bounded. Therefore the plan S' where $S'(\gamma) = S_{k-1}(\gamma)$ for $\gamma \neq \gamma_k$ and $S'(\gamma_k) = 1$ is a bounded plan and hence contains more than n b.ps. Thus S_{k-1} contains more than $n+1$ b.ps. and hence so does S_0.□

Most of the material in this section is due to B. Brainerd, whose Theorems 2A and 2B give a very elegant characterization of simple s.ps. Indeed, by combining them, we obtain the following theorem combinatorially.

THEOREM 2C (DeGroot) *A necessary and sufficient condition for a s.p. of size n to be simple is that it have exactly $n+1$ b.ps.*

We conclude with two examples which clarify the concepts of non-admissible deformations and a suitable choice of a series of admissible deformations.

EXAMPLES Consider the non-simple sampling plan defined by the figure below. The plan has an essential boundary, although examination of the line $x+y=4$ shows it is non-simple.

	y	·	·	·	·	·	·	·	
	·	x^{γ_2}	·	·	·	·	·	·	
	x^{γ_1}	·	x^{γ_3}	·	·	·	·	·	
	·	·	x^{γ_4}	·	·	·	·	·	
	·	·	·	x^{γ_5}	x^{γ_6}	·	·	·	
0	·	·	·	·	·	x^{γ_7}	·	·	x

Clearly S is defined by its b.ps. $\gamma_1, \gamma_2, \gamma_3, \gamma_4, \gamma_5, \gamma_6$, and γ_7. The deformation of S at γ_3 is not bounded, and the deformation at γ_4 has γ_5 as its non-essential b.p. These are therefore non-admissible deformations of S. The proof of Theorem 2B may be illustrated by such a figure.

To illustrate Theorem 2A, consider the figure below representing a simple s.p. of size 5:

	y	·	·	·	·	·	·	·	
	·	·	·	·	·	·	·	·	
	·	x^{γ_2}	x^{γ_3}	·	·	·	·	·	
	·	·	·	x^{γ_4}	·	·	·	·	
	x^{γ_1}	·	·	x^{γ_5}	·	·	·	·	
0	·	·	·	·	x^{γ_6}	·	·	·	x

We suggest the following series of admissible deformations:

S_1: γ_1 shifted to (0, 2);
S_2: γ_1 shifted to (0, 3);
S_3: γ_1 shifted to (0, 4);
S_4: γ_2 shifted to (1,4);
S_5: γ_1 shifted to (0, 5);
S_6: γ_5 shifted to (4, 1);
S_7: γ_6 shifted to (5, 0).

Evidently S_7 has all b.ps. of index n. In this process, γ_1 was pushed upwards four times, γ_2 once, γ_3 and γ_4 remained in their respective positions, and γ_5 and γ_6 were pushed once each to the right. The number of times any b.p. is pushed to coincide with the corresponding b.p. of the single s.p. of the same size essentially defines the characterization that we propose to discuss. For example, according to our characterization the simple s.p. S corresponds to the vector (4, 1, 0, 0, 1, 1).

2.5. *Enumeration of simple s.ps.*

As the connection between Young chains and admissible deformations should now be clear from our examples and Theorem 2A, we can describe characterizations of special classes of simple s.ps. with very informal proofs in this section. In every case it is convenient to conceive the b.ps. of the single s.p. of the same size as being moved inwards, starting first with those on the y-axis and x-axis, moving next to the lines $x = 1$ and $y = 1$, etc., till the required s.p. is attained. Lemmas 2D and 2E below are concerned with symmetric plans, i.e. plans symmetric about the line $x = y$, where symmetric deformations on both the x-axis and y-axis, etc. are performed simultaneously.

LEMMA 2D *A symmetric s.p. of size* $2n$ *is characterized by symmetric vectors*

$$(d_{n-1}, d_{n-2}, \ldots, d_1, 0, 0, 0, d_1, \ldots, d_{n-2}, d_{n-1})$$

where

[2j] $\quad 0 \leq d_1 \leq \ldots \leq d_{n-1}$ *and* $d_i \leq 2i \quad (i = 1, \ldots, n-1)$.

The number of such plans is $\frac{1}{n}\binom{3n}{n-1}$.

□Let A, B be two vectors satisfying [2j] and $A \,\mathrm{D}\, B$. Then clearly the s.p. corresponding to A is contained in the s.p. corresponding to B, perfectly illustrating the dominance relation and Young chains. An application of Lemma 3B, Chapter I, yields the enumeration.□

LEMMA 2E *The number of symmetric s.ps. of size* $2n+1$ *is* $\frac{1}{n+1}\binom{3n+1}{n}$, $n = 0, 1, 2, \ldots$

□The proof is very similar to that of Lemma 2D.□

We present now an example which shows that there exists a 1:1 correspondence between $S_{2n}{}^*$, the simple symmetric plans of size $2n$ of Lemma 2D, and simple s.ps. of size n. Consider the vector

[2k] $\quad v_8 = (1, 4, 4, 5, 8, 10, 13) \in S_{16}{}^*$.

The closest vector with even elements dominating [2k] is clearly

[2l] $e_8 = (2, 4, 4, 6, 8, 10, 14)$.

Construct $e_8 - v_8 = (1, 0, 0, 1, 0, 0, 1)$ and $\frac{1}{2}e_8 = (1, 2, 2, 3, 4, 5, 7)$. Partition $\frac{1}{2}e_8$ into two vectors, one non-increasing and the other non-decreasing, according to the zero and non-zero elements of $e_8 - v_8$. The two vectors required are

[2m] $v_8^1 = (5, 4, 2, 2)$, $\quad v_8^2 = (1, 3, 7)$.

Form the vector $A_8 = (v_8^1, 0, 0, v_8^2)$, yielding

[2n] $A_8 = (5, 4, 2, 2, 0, 0, 1, 3, 7)$.

Then A_8 represents a simple s.p. of size n, which is in 1:1 correspondence with $v_8 \in S_{2n}^*$.

LEMMA 2F *There exists a* 1:1 *correspondence between the simple symmetric s.ps. of size* $2n$, *i.e.* S_{2n}^*, *and the simple s.ps. of size* n, S_n *(say).*

□See exercises 11 and 12.□

Table 7 gives the vectors in S_n, i.e. the boundary deformation vectors of simple s.ps. for $n = 1, 2, 3, 4$. Many amusing combinatorial identities can be derived from our lemmas and an inspection of Table 7, of which we indicate one yielding a very simple practical check on vectors $A_n \in S_n$. Let $W(A_n)$ denote, as usual, the sum of elements in A_n, e.g. $W(A_8)$ in [2n] is 24. A s.p. A_n is said to be *minimal* if $W(A_n) = \frac{1}{2}n(n-1)$. We also note, in passing, that any vector $A_n \in S_n$ has the structure given by [2m], i.e. a non-increasing part A_n^1 (possibly empty), then two zeros, and a non-decreasing part A_n^2 (possibly empty).

LEMMA 2G *The number of minimal sampling plans of size* n *is* 2^{n-1} $(n = 1, 2, \ldots)$.

□Partition the set $(1, 2, \ldots, n-1)$ into two subsets A_n^1 (which is non-increasing) and A_n^2 (which is non-decreasing). The minimal plans are given by $(A_n^1, 0, 0, A_n^2)$. □

A check for simple s.ps. of size n may be obtained from Lemma 2G, since given any vector A_n supposed to belong to S_n, there exists through dominance considerations a minimal plan of size n at least as small as the given plan A_n.

For a deep discussion on s.ps. both finite and infinite, we refer to Kagan, Linnik, and Rao (1973); for further combinatorial aspects see Narayana and Mohanty (1963).

TABLE 7

The vectors in S_n ($n = 1, 2, 3, 4$)

n	S_n
1	(0, 0)
2	(0, 0, 0)
	(0, 0, 1)
	(1, 0, 0)
3	(0, 0, 0, 0)
	(0, 0, 0, 1)
	(0, 0, 0, 2)
	(1, 0, 0, 0)
	(2, 0, 0, 0)
	(0, 0, 1, 1)
	(0, 0, 1, 2)
	(1, 0, 0, 1)
	(1, 0, 0, 2)
	(2, 0, 0, 1)
	(1, 1, 0, 0)
	(2, 1, 0, 0)
4	(0, 0, 0, 0, 0)
	(0, 0, 0, 0, 1)
	(0, 0, 0, 0, 2)
	(0, 0, 0, 0, 3)
	(1, 0, 0, 0, 0)
	(2, 0, 0, 0, 0)
	(3, 0, 0, 0, 0)
	(0, 0, 0, 1, 1)
	(0, 0, 0, 1, 2)
	(0, 0, 0, 1, 3)
	(0, 0, 0, 2, 2)
	(0, 0, 0, 2, 3)
	(1, 0, 0, 0, 1)
	(1, 0, 0, 0, 2)
	(1, 0, 0, 0, 3)
	(2, 0, 0, 0, 1)
	(2, 0, 0, 0, 2)
	(2, 0, 0, 0, 3)
4 (cont'd)	(3, 0, 0, 0, 1)
	(3, 0, 0, 0, 2)
	(1, 1, 0, 0, 0)
	(2, 1, 0, 0, 0)
	(3, 1, 0, 0, 0)
	(2, 2, 0, 0, 0)
	(3, 2, 0, 0, 0)
	(0, 0, 1, 1, 1)
	(0, 0, 1, 1, 2)
	(0, 0, 1, 1, 3)
	(0, 0, 1, 2, 2)
	(0, 0, 1, 2, 3)
	(1, 0, 0, 1, 1)
	(1, 0, 0, 1, 2)
	(1, 0, 0, 1, 3)
	(1, 0, 0, 2, 2)
	(1, 0, 0, 2, 3)
	(2, 0, 0, 1, 1)
	(2, 0, 0, 1, 2)
	(2, 0, 0, 1, 3)
	(3, 0, 0, 1, 1)
	(3, 0, 0, 1, 2)
	(1, 1, 0, 0, 1)
	(2, 1, 0, 0, 1)
	(3, 1, 0, 0, 1)
	(2, 2, 0, 0, 1)
	(3, 2, 0, 0, 1)
	(1, 1, 0, 0, 2)
	(2, 1, 0, 0, 2)
	(3, 1, 0, 0, 2)
	(1, 1, 0, 0, 3)
	(2, 1, 0, 0, 3)
	(1, 1, 1, 0, 0)
	(2, 1, 1, 0, 0)
	(3, 1, 1, 0, 0)
	(2, 2, 1, 0, 0)
	(3, 2, 1, 0, 0)

SOURCE: Prepared by Dr S. G. Mohanty.

EXERCISES

1 Let $A = (a_1, \ldots, a_n)$ be a path to (m, n) with horizontal ranks $r_1, \ldots, r_m$. Show that

$$W(A) = \sum_{i=1}^{m} r_i - \frac{m(m+1)}{2}.$$

Thus the U-test and Wilcoxon test are essentially identical.

2 Prove Lemma 2A by introducing a Young chain between A and B.

3 Given that the UMP test (against Lehmann alternatives) at level $\alpha = 0.1$ with $k = 2$, $m = n = 6$, consists of the union A of GF regions listed below, verify that A contains 92 paths.

$$A = 25 \cup 115 \cup 44 \cup 234 \cup 11134 \cup 2224 \cup 11224 \cup 111124 \cup 2333 \cup 111333 \cup 22233 \cup 112233 \cup 222223.$$

(The 93rd path is 111224.)

4 Establish formula [4d]. Extend it to $L_1, L_2, \ldots, L_n$.

5 Using exercise 4, verify all the levels in Table 2. Also verify that the regions satisfy [2f].

6 Show that the powers of the Wilcoxon test, UMP test, and $DL_2(2)$ test as given by Table 2 (when randomized for the exact level) yield 0.3796, 0.4060, and 0.3999 respectively when $m = n = 6$, $\alpha = 0.1$. [A computer is needed for the heavy calculations, even noting that the power of the Wilcoxon test can be obtained from Shorack (1967).]

7 Show that the Wilcoxon and dominance ratio tests are dominance tests. Show further that both tests are *self-dual* tests, in the sense that when $m = n$ every path and the dual path are either both included in or both excluded from the critical region.

8 The normal scores test associates with a path with ranks $r_1, \ldots, r_m$ the sum of scores

$$\Phi^{-1}\left(\frac{r_1}{N+1}\right) + \ldots + \Phi^{-1}\left(\frac{r_m}{N+1}\right)$$

where $N = m+n$ and Φ is the cumulative distribution function of the normal distribution. Show that the test based on such scores is also a self-dual dominance test. (See Hájek's (1969) Van der Waerden test.)

9 Verify from Table 6 when $n = 3$, $m = 9$ that the 5% critical regions of the Wilcoxon test and DR test are (respectively)

$112 \cup 22 \cup 13 \cup 4$ and $111 \cup 22 \cup 13 \cup 5$.

†10 (Conjecture) Does every region for the UMP test against Lehmann alternatives include a path in the U series defined after [2e]? If so, does this path have maximal area underneath it, as compared to other paths defining the critical region? (In exercise 3 such a path is 222223.)

11 Let $n, x_i, y_i\,(i = 1, \ldots, k)$ be non-negative integers. For $n \geq 1$, consider the function analogous to the multinomial defined recursively as follows:

$$(n{:}x_1, \ldots, x_k) = \begin{cases} 0 & \text{if } \sum_{i=1}^{n} x_i > n, \\ \sum_{y_1=0}^{x_1} \cdots \sum_{y_k=0}^{x_k} (n-1; y_1, \ldots, y_k) & \text{otherwise,} \end{cases}$$

where $(0; 0, \ldots, 0) = 1$. Prove that

$$n;\, x_1, \ldots, x_k) = \prod_{i=1}^{k} \binom{n+x_i}{x_i} \left[1 - \frac{\sum x_i}{n+1}\right] \quad \text{for } \sum x_i \leq n.$$

12 (11 continued) Let $A_{n,k}$ denote the set of vectors $(a_1, \ldots, a_n)$ dominated by $(k, 2k, \ldots, nk)$, where $n, k \geq 2$ are positive integers. Let $[n; x_1, \ldots, x_k]$ be the subset of $A_{n,k}$ such that every vector in $[n; x_1, \ldots, x_k]$ has exactly x_i of its *positive* elements congruent to $i \pmod k$, $i = 1, \ldots, k$. Then

$$|[n; x_1, \ldots, x_k]| = (n; x_1, \ldots, x_k).$$

(a) When $k = 2$, show that this result illustrates the 1:1 correspondence between $S_{2n}{}^*$ and S_n.

(b) Extend the above result to vectors dominated by $(k+d, \ldots, nk+d)$, where $0 \leq d < k$, and a suitable generalization of $(n; x_1, \ldots, x_k)$. Obtain thus a refinement of ballot theorems and Lemma 3B in Chapter I.

†13 Consider the class of dominance tests $D = \{c_1, W, DR(a)\}$ where $DR(a)$ for $a \geq 0$ represents the dominance test (generalizing [2i])

$$DR_a(A) = V^a(A)/V(m - A')$$

where $A' = (a_n, \ldots, a_1)$ is the transpose of A. Verify that the Wilcoxon test is inadmissible in the class D for the range of Milton's Table 1 (reference 8 in bibliography at end of book). Specifically, show that in Milton's Table 1, p. 30: (i) $DR(1/2) = W$ for $(m, n) = (7, 6)$, $0.2 \leq d \leq 3.0$; (ii) $DR(3/4) > W$ for $(m, n) = (7, 5)$, $d = 3.0$; (iii) $DR(1) \geq W$, or $D(1/2) \geq W$, for all other cases where $W > c_1$. (*Note*: This combinatorial inadmissibility extends the usual concept of inadmissibility due to Wald.)

REFERENCES

1 BRAUER, A., I. GENTRY, and K. SHAW. 1968. 'A new proof of a theorem by H. G. Landau on tournament matrices,' *J. Comb. Theory 5*, 289–92
2 DEGROOT, M. H. 1959. 'Unbiased sequential estimation for binomial populations,' *Ann. Math. Statist. 30*, 80–101
3 FELLER, W. 1968. *An Introduction to Probability Theory and Its Applications*, Vol. I. John Wiley and Sons, New York
4 GIRSHICK, M. A., F. MOSTELLER, and L. J. SAVAGE. 1946. 'Unbiased estimates for certain binomial sampling plans with applications,' *Ann. Math. Statist. 17*, 13–23
5 HÁJEK, J. 1969. *A Course in Nonparametric Statistics*. Holden-Day, San Francisco
6 KAGAN, A. M., YU. V. LINNIK, and C. R. RAO. 1973. *Characterization Problems in Mathematical Statistics*. John Wiley and Sons, New York
7 KREWERAS, G. 1967. 'Traitement simultané du "Problème de Young" et du "Problème de Simon Newcomb,"' *Cahiers du Bur. Univ. de Rech. Opér. 10*, 23–31
8 LANDAU, H. G. 1953. 'On dominance relations and the structure of animal societies: III,' *Bull. Math. Biophys. 15*, 143–8
9 LEHMANN, E. L. 1953. 'The power of rank tests,' *Ann. Math. Statist. 24*, 23–43
10 LEHMANN, E. L., and C. STEIN. 1950. 'Completeness in the sequential case,' *Ann. Math. Statist. 21*, 376–85
11 MOON, J. W. 1968. *Topics in Tournaments*. Holt, Rinehart and Winston, New York
12 NARAYANA, T. V. 1955. 'Sur les treillis formés par les partitions d'un entier; leurs applications à la théorie des probabilités,' *Comp. Rend. Acad. Sci. Paris 240*, 1188–9
13 NARAYANA, T. V., and S. G. MOHANTY. 1963. 'Some properties of compositions and their application to probability and statistics II.' *Biom. Z. 5*, 8–18
14 SHORACK, R. A. 1967. 'Tables of the distribution of the Mann-Whitney-Wilcoxon U-Statistic under Lehmann alternatives,' *Technometrics 9*, 666–77
15 STECK, G. P. 1969. 'The Smirnov tests as rank tests,' *Ann. Math. Statist. 40*, 1449–66
16 – 1974. 'A new formula for $P(R_i \leq b_i, 1 \leq i \leq m \mid m, n, F = G^k)$,' *Ann. Prob. 2*, 155–60

IV

The combinatorics of knock-out tournaments

Ars longa, vita brevis.

Ach Gott! die Kunst ist lang;
Und kurz ist unser Leben

Goethe, *Faust*

The lyfe so short, the crafte so long to learn.

Chaucer, Prologue

In this chapter we shall apparently digress far from lattice path combinatorics as we study a class of random knock-out tournaments following Narayana (1968) and Narayana and Zidek (1969). However, dominance still plays a certain role in such tournaments, and as will be seen in Chapter V, the random walk (lattice path) aspect is crucial for our applications.

The combinatorics of repeated round-robin tournaments has been studied in detail by statisticians interested in paired comparisons, as well as by mathematicians as a branch of graph theory. Although it has long been surmised that repeated knock-out tournaments could represent an alternative to repeated round-robin tournaments, very few results on the general knock-out tournament were available until recently. A convenient summary of known results on round robins and the classical knock-out involving $n = 2^t$ players is available in the books of David (1963) and Moon (1968), who provide many references. We introduce a class of random knock-out tournaments and study their elementary combinatorial properties before applying these results to selection problems in the next chapter.

The basic model to which we restrict ourselves has been discussed by David (1963). Just one of the players, A say, has a probability p of winning against each of the other $n-1$ players B_i ($i = 1, \ldots, n-1$), who are of equal strength. We shall assume, implicitly, that $p \geq \frac{1}{2}$, although most, if not all of our formulae are valid for all p, $0 \leq p \leq 1$. This assumption expresses in simplified form the idea that a superior player is present. It is relevant to point out that, apart from the mathematical tractability of this model, it represents the important case of a single outlier. In comparing selection procedures for choosing the best object, this model provides a suitable basis for comparison; we assume the reader is familiar with David's book quoted above, particularly Chapter VI where selection problems involving this and other models are discussed.

After providing a suitable motivation by discussing in detail the case $n = 2^t$, we define in Section 2 a class of random knock-out tournaments. No effort is made to study in this chapter the combinatorial aspects of *repeated* knock-outs; only basic properties of single replications are investigated. In Section 3, the single replication cases of various knock-out tournaments are compared, giving us important clues regarding the design of repeated knock-out tournaments. We conclude with an asymptotic result which arises naturally from our combinatorial approach, and describe a knock-out procedure with some balance which has the best features of both round-robins and knock-outs.

1
THE CLASSICAL CASE

A widely known elementary result is sometimes posed as an exercise in introductory textbooks on probability (cf. Tucker 1962, p. 18). $n = 2^t$ players of equal strength play a knock-out tournament, being randomly paired off in each round. What is the probability that two given players A and B meet in some round of the tournament? The answer $2/n$ could be obtained by noting that $n-1$ matches are played and any one of the $\binom{n}{2}$ pairs formed from the n players is likely to play in a match. Indeed the basic idea is that a knock-out tournament has $n-1$ matches, i.e. $n-1$ losers and a winner, each loser requiring one match to be 'knocked-out.' Thus n need not be a power of 2 exactly, and the answer $2/n$ is valid for all integral $n \geq 2$.

More generally, when $n = 2^t$ players of equal strength are randomly playing a knock-out, let $R_n{}^i$ denote the probability that a given player A plays exactly i rounds in this tournament. Clearly $R_n{}^i = 0$ for $i > t$. As a preliminary to applying the inclusion-exclusion principle, we denote by $P_n^{[i]}$ and P_n^i, respectively, the probabilities that A meets only $B_1, \ldots, B_i$ and that A meets $B_1, \ldots, B_i$ and perhaps others in the tournament. Clearly, from the geometric distribution (Feller 1968),

$$[1a] \quad R_n^i = \frac{1}{2^i} \quad (i < t), \qquad R_n^t = \frac{1}{2^{t-1}}.$$

$P_n{}^{[i]}$ is immediately obtained from the $R_n{}^i$; indeed, as the i opponents of A are equally likely to be any one of the $\binom{n-1}{i}$ combinations, we have

[1b] $P_n^{[i]} = \dfrac{1}{2^i\binom{n-1}{i}} \quad (i<t), \qquad P_n^{[t]} = \dfrac{1}{2^{t-1}\binom{n-1}{t}}.$

The P_n^i are now obtained from the $P_n^{[i]}$ by analogy with the inclusion and exclusion principle. Alternatively, if A meets $B_1, \ldots, B_i$ and v additional players, these v additional players can be chosen in $\binom{n-i-1}{v}$ ways from the remaining players ($v = 0, 1, \ldots, t-i$). Noting that these cases are mutually exclusive, an appropriate application of the theorem of total probabilities yields

[1c] $P_n^i = \sum_{v=0}^{t-i} P_n^{[i+v]}\binom{n-i-1}{v}.$

Substituting from [1b] into [1c], after an elementary simplification we have the relation connecting P_n^i and R_n^i, namely

[1d] $P_n^i = \dfrac{1}{\binom{n-1}{i}} \sum_{v=0}^{t-i} R_n^{i+v}\binom{i+v}{i},$

where, by definition, $R_n^i = P_n^i = 0$ for $i>t$ in the classical case. Clearly the equations [1d] may be considered as n simultaneous equations for the P_n^i in terms of the R_n^i. Inversely, given the P_n^i, if the R_n^i were unknown, we could solve for the R_n^i. Indeed the inverse of the matrix of coefficients of R_n^i in [1d] is well known (see Feller (1968) for a similar proof) and we obtain

[1e] $R_n^i = \sum_{v=0}^{n-i-1} (-1)^v P_n^{i+v}\binom{n-1}{i+v}\binom{i+v}{i}.$

Of course $R_n^i = P_n^i = 0$ for $i>t$ in the classical case. We have stated equations [1d], [1e] generally, since they may be validated for all random tournaments of our basic model, as defined in the next section.

We turn now to the classical case where A has probability $p \neq \frac{1}{2}$ of defeating any B_i. Since the B_i's are of equal strength, it is evident that equations [1d], [1e] continue to hold by the same 'symmetry' arguments as used before. (The reader interested in a purely combinatorial derivation should consult Capell and Narayana (1970), particularly if he is unconvinced by symmetry arguments.)

Hence

[1f] $R_n^i = p^{i-1}q$ for $i = 1, \ldots, t-1, \qquad R_n^t = p^{t-1}.$

Substituting these values in [1d], we have, in the classical case with general p,

[1g] $\binom{n-1}{i}P_n^i = x_i + p^t\binom{t}{i},$

where

[1h] $x_i = \sum_{v=0}^{t-i} p^{i+v-1} q \binom{i+v}{i}, \qquad i = 1, \ldots, t.$

A little simplification gives

[1i] $x_1 = \frac{1-p^t}{q} - tp^t \quad \text{and} \quad x_i = \frac{p}{q} x_{i-1} - \binom{t}{i-1} p^t - p^t \binom{t}{i}.$

Recursion [1i] and induction yield

[1j] $$x_i = \frac{1}{q^i}\left\{p^{i-1}(1-p^t) - p^t\left[\binom{t}{1} q p^{i-2} + \binom{t}{2} q^2 p^{i-3} + \ldots + \binom{t}{i-1} q^{i-1}\right]\right\} - \binom{t}{i} p^t.$$

Thus, from [1g] and [1j], we have explicit expressions for $P_n^{\,i}$:

[1k] $$\binom{n-1}{i} P_n^i = \left(\frac{p}{q}\right)^{i-1} \frac{p^t}{q}\left[\frac{1}{p^t} - \binom{t}{0} - \binom{t}{1}\frac{q}{p} - \ldots - \binom{t}{i-1}\left(\frac{q}{p}\right)^{i-1}\right] \qquad (i = 1, \ldots, t).$$

When $p = q$, the expression for P_n^i simplifies, using $n = 2^t$. We then have

$$\binom{n-1}{i} P_n^{\,i} = \frac{2}{n}\left[n - \binom{t}{0} - \binom{t}{1} - \ldots - \binom{t}{i-1}\right].$$

In particular, from [1k],

[1l] $P_n^1 = (1-p^t)/q(n-1),$

which naturally reduces to $2/2^t = 2/n$ when $p = \frac{1}{2}$.

As a conclusion to this section, we make one further remark about P_n^1 which will be valid for random tournaments in general. Clearly $E(R)$, the expected number of rounds that A plays in the tournament, equals $(n-1)P_n^1$. Letting $\Pi = p^t$ be the probability that A wins the tournament, we obtain as a reformulation of [1l]

$$\Pi + qE(R) = 1.$$

2
RANDOM KNOCK-OUT TOURNAMENTS

When n is not a power of 2, say $n = 2^t + K, 0 < K < 2^t$, it is usual in the literature to reduce the number of players to 2^t by matching $2K$ players randomly in a preliminary round. Motivated by this idea, we define for every integer $n \geq 2$ a random knock-out tournament with n players as a vector of positive integers $(m_1, \ldots, m_k)$ satisfying

$$[2a] \quad \begin{cases} m_1 + \ldots + m_k = n-1, \quad m_k = 1; \\ 2m_1 \leq n; \\ 2m_i \leq n - m_1 - \ldots - m_{i-1} \qquad (i \geq 2). \end{cases}$$

A tournament defined by the vector $(m_1, \ldots, m_k)$ is played as follows. On the first round $2m_1$ players, chosen at random from n, are paired off randomly. The remaining $n-2m_1$ players have a bye for this round. The m_1 pairs yield m_1 losers who are eliminated from the tournament. We are then left with a tournament of $n-m_1$ players, with vector $(m_2, \ldots, m_k)$. This inductive rule is well defined for $n > 2$, since in the case $n = 2$ there is a unique tournament of one round.

We indicate here a few examples of tournaments which have been studied in the literature and are of special importance in the applications. For any $n = 2^t + K$ $(0 \leq K < 2^t)$ we designate by

T_1: the tournament with vector $m_i = [(n+2^{i-1}-1)/2^i]$,

T_2: the tournament with vector $(K, 2^{t-1}, 2^{t-2}, \ldots, 1)$,

T_3: the tournament with vector $(1, \ldots, 1)$.

When $K = 0$ both T_1 and T_2 reduce to the classical case; but if $K \neq 0$ T_1 and T_2 consist of $t+1$ rounds. Of course, T_3 always consists of $n-1$ rounds. It is convenient to set

$$[2b] \quad n_1 = n, \qquad n_i = n_{i-1} - m_{i-1} \quad (i \geq 2),$$

so that n_i is the number of players at the start of round i $(i = 1, \ldots, k)$. Further, let

$$[2c] \quad p_i = 2m_i/n_i, \quad q_i = 1 - p_i \qquad (i = 1, \ldots, k),$$

so that p_i is the probability that a specified player, among the n_i qualified for round i, does *not* get a bye in the round. Finally, from [2a], $n_k = 2$ so that $p_k = 1$.

We emphasize once again that we are studying the model where one player A has the probability p of winning in a match against each of the other equally strong $B_1, \ldots, B_{n-1}$. Let $\Pi = \Pi(n; m_1, \ldots, m_k)$ denote the probability that A wins the tournament $(m_1, \ldots, m_k)$ where the m_i satisfy [2a]. Also, let us suppose we are given the vector

$$R_n = (R_n^1, \ldots, R_n^k) = R(n; m_1, \ldots, m_k)$$

where R_n^i represents the probability that A plays exactly i rounds. As $\sum_i R_n^i = 1$, $E(R_n) = \sum_i iR_n^i$ represents the expected number of rounds played by A in the tournament. The above definitions lead us to the following lemma.

LEMMA 2A *The probability* $\Pi = \Pi_n$ *that A wins the tournament* $(m_1, \ldots, m_k)$ *is given by*

$$[2d] \quad \Pi = (p_1 p + q_1)(p_2 p + q_2) \ldots (p_{k-1} p + q_{k-1}) p.$$

Further

$$[2e] \quad \Pi + qE(R_n) = 1, \quad \textit{where } q = 1 - p.$$

□We prove both results by induction. Let us assume that [2d] is true for *all* tournaments with $n-1$ or fewer players, since it is clearly true for all tournaments with $n = 2, 3, 4$ players. The inductive hypothesis assures us that

$$[2f] \quad \Pi_{n_2} = \Pi(n_2; m_2, \ldots, m_k) = (p_2 p + q_2) \ldots (p_{k-1} p + q_{k-1}) p$$

where $p_2, q_2, \ldots, p_{k-1}, q_{k-1}$ are defined by [2c]. Now player A can survive the first round of $(m_1, \ldots, m_k)$ in two exclusive and exhaustive ways, namely either by playing and winning round 1 with probability pp_1 or by having a bye with probability q_1. An elementary theorem of probability gives

$$[2g] \quad \Pi_n = (p_1 p + q_1) \Pi_{n_2}.$$

A similar inductive proof establishes [2e] with the help of the recursion similar to [2g]:

$$E(R_{n_1}) = p_1 q + p_1 p[E(R_{n_2}) + 1] + q_1 E(R_{n_2}).$$

Of course $E(R_{n_2})$ clearly refers to the tournament $(m_2, \ldots, m_k)$.□

We restate [2d] informally as follows: Let us suppose we multiply out the right-hand side of [2d], yielding

$$[2h] \quad \Pi = b_{k-1} p + b_{k-2} p^2 + \ldots + b_0 p^k = \mathbf{b}'\mathbf{w}',$$

where

$$[2i] \quad \mathbf{b}' = (b_{k-1}, \ldots, b_0) \quad \text{and} \quad \mathbf{w}' = (p, p^2, \ldots, p^k).$$

Now, given that A wins the tournament, let B_{k-i} be the probability that A receives exactly $k-i$ byes in the course of the tournament ($i = 1, \ldots, k$). Since the conditional probability that A wins the tournament given that he has received $k-i$ byes is p^i, we have

$$[2j] \quad \Pi = B_{k-1} p + \ldots + B_0 p^k.$$

Comparing [2h] and [2j], we see that $B_{k-i} \equiv b_{k-i}$ ($i = 1, \ldots, k$), so that b_{k-i}, as defined by [2h], is indeed the conditional probability that A receives exactly $k-i$ byes given that A won the tournament. By an exactly analogous argument, R_n^i may be split up into the exclusive probabilities that A plays i rounds and wins the tournament or that A plays i rounds and loses. Since the details are obvious, we state the following lemma.

LEMMA 2B *Let* $\mathbf{c} = (c_{k-1}, \ldots, c_0)$ *where*

$$c_{k-1} = 1, \qquad c_{k-i} = 1 - \sum_{j=1}^{i-1} b_{k-j} \quad (i \geq 2).$$

Then

$$R_n^i = (p^i b_{k-i} + p^{i-1} q\, c_{k-i}).$$

Finally, let p_n^i and Q_n^i denote respectively the probabilities that A meets $B_1, B_2, \ldots$, and B_i and (B_1 or B_2 or……B_i) in the tournament. By the usual arguments involving inclusion and exclusion, the following relations hold:

THEOREM 2A

$$P_n^i = \frac{1}{\binom{n-1}{i}} \sum_{v=0}^{n-i-1} R_n^{i+v} \binom{i+v}{i}, \qquad Q_n^i = \sum_{v=1}^{n-1} (-1)^{v-1} P_n^v \binom{i}{v};$$

the reciprocal relations are also valid:

$$R_n^i = \sum_{v=0}^{n-i-1} (-1)^v P_n^{i+v} \binom{n-1}{i+v} \binom{i+v}{i}, \qquad P_n^i = \sum_{v=1}^{n-1} (-1)^{v-1} Q_n^v \binom{i}{v}.$$

Remembering that R_n^i, P_n^i are zero for $i > k$ for any tournament $(m_1, \ldots, m_k)$, we can obtain numerical values for P_n^i and Q_n^i by using a computer for quite large (1000) values of n. Before applying Theorem 2A to obtain explicit results, we introduce the important 'play the winner' tournament T_4 which is of particular importance in statistical applications. For any $n \geq 2$, the tournament T_4 is played according to the vector $(1, 1, \ldots, 1)$ with the further restriction that the winner of any round automatically plays in the succeeding round. Thus, after a pair, drawn at random, plays in the first round, one of the remaining $n-2$ bye players of round 1 meets the winner in round 2, and so on. We could further generalize our definition of random knock-out tournaments [2a] following the play-the-winner idea of T_4 but do not do so. Clearly Lemma 2B and Theorem 2A apply to T_4, whereas [2d] does not. Applying Theorem 2A to the tournaments T_1, T_2, T_3, T_4 for the non-trivial cases $n > 4$, where, we recall, $n = 2^t + K$ with $0 \leq K < 2^t$, we summarize explicitly or in recurrence form values for R_n^i, P_n^i for tournaments T_1, T_2, T_3, T_4.

Tournament T_1

$$R_{2n}^1 = q, \qquad R_{2n}^{i+1} = pR_n^i \quad (i = 1, \ldots, t).$$

For $i>2$, after some simplification

$$P^i_{2n} = \frac{p\binom{n-1}{i}}{\binom{2n-1}{i}}\left[P^i_n + \frac{i}{n-i}P^{i-1}_n\right].$$

A similar argument yields, for $i>2$,

$$P^i_{2n-1} = \frac{\binom{n-1}{i}}{\binom{2n-1}{i}}\left[\frac{pi}{n-i}\frac{n-1}{n}P^{i-1}_n + \frac{\frac{1}{2}+p(n-1)}{n}P^i_n\right].$$

Tournament T_2

$$\Pi = \frac{2K}{2^t+K}p^{t+1} + \frac{2^t-K}{2^t+K}p^t.$$

$$R^i_n = p^{i-1}q \qquad (i=1,\ldots,t-1);$$

$$R^t_n = \frac{2K}{n}p^{t-1}q + \frac{2^t-K}{n}p^{t-1};$$

$$R^{t+1}_n = \frac{2K}{n}p^t.$$

$$\binom{n-1}{i}P^i_n = \left(\frac{p}{q}\right)^{i-1}\frac{p^t}{q}\left\{\frac{1}{p^t} - \binom{t}{0} - \frac{q}{p}\binom{t}{1} - \cdots - \left(\frac{q}{p}\right)^{i-1}\binom{t}{i-1}\left[\frac{1-2Kq}{n}\right]\right\}.$$

Tournament T_3

$$\Pi = \frac{(2p+n-2)_{n-1}}{n!}.$$

$$R^{i+1}_n = \frac{2}{n}pR^i_{n-1} + \left(1-\frac{2}{n}\right)R^{i+1}_{n-1} \qquad (i=1,\ldots,n-2);$$

$$R^1_n = q.$$

Explicit results for P^i_n appear to be cumbersome (if not impossible), but numerical results are easily obtained with a computer.

Tournament T_4
Explicit results are possible in this case.

$$\Pi = \frac{1}{n}p + \frac{1}{n}p^2 + \dots + \frac{1}{n}p^{n-2} + \frac{2}{n}p^{n-1}$$

$$= \frac{1}{n}p\left(\frac{1-p^{n-1}}{1-p}\right) + \frac{1}{n}p^{n-1}.$$

$$R_n^i = p^{i-1}q\frac{n-1}{n} + \frac{p^{i-1}}{n} \quad (i<n-1), \qquad R_n^{n-1} = p^{n-2}\frac{2}{n}.$$

$$P_n^i = \frac{p^{i-1}}{n\binom{n-1}{i}} \sum_{v=0}^{n-2-i} \overline{(qn-i-v+1)}\binom{i+v}{i}p^v + p^{n-2}\frac{2}{n}.$$

3
A COMPARISON OF TOURNAMENTS

In this section we shall assume that $\frac{1}{2} \leq p \leq 1$, so that a stronger player A is present. We let $\Pi_i(p)$ $(i = 1, 2, 3, 4)$ be the probability that A wins the corresponding tournament T_i, and state our result for all $n \geq 4$, since for $n = 2, 3$ only one tournament is possible. Many results are true even if $p < \frac{1}{2}$.

THEOREM 3A *For any tournament with $n \geq 4$ players, and any (fixed) p lying in the open interval $(\frac{1}{2}, 1)$,*

$$\Pi_2(p) \geq \Pi_1(p) > \Pi_3(p) > \Pi_4(p).$$

As the details of the proof are quite elementary, though tedious, we restrict ourselves to stating the definition of descendant of a tournament, which is a useful tool in comparing tournaments.

DEFINITION A Let $T = (m_1, \dots, m_k)$ be a tournament vector for n players, and for some $i < k$, let $m_i = m_i' + m_i''$, where m_i', m_i'' are positive integers. Then the tournament with vector $T' = (m_1, \dots, m_{i-1}, m_i', m_i'', m_{i+1}, \dots, m_k)$ will be called a first-stage *descendant* of T.

REMARK From Lemma 2A, it is immediately verified that $\Pi(T) > \Pi(T')$ for fixed p in $(\frac{1}{2}, 1)$. Since it is fairly obvious how to define second-stage (and further) descendants of T, and that $(1, 1, \dots, 1)$ is a descendant of every random knock-out tournament, one part of our theorem follows easily. Also W. Maurer (1975) has given recently a more general version of one part of Theorem 3A. For these reasons we include only a sketch of the proof of Theorem 3A.

SKETCH OF PROOF (for fuller details see Hopkins (1969))

(i) $\Pi_3(n) > \Pi_4(n)$ for all $n \geq 4$. This is true for $n = 4$, and induction completes the proof.

(ii) $\Pi_3(n) < \Pi_1(n)$; $\Pi_3(n) < \Pi_2(n)$ for $n \geq 4$. This follows from the remark that if T' is a first-stage descendant of T, then $\Pi_T > \Pi_{T'}$, for $\frac{1}{2} < p < 1$ and any fixed $n \geq 4$.

(iii) $\Pi_2(n) \geq \Pi_1(n)$. It is easy to prove, using T_2, that A has a probability $\Pi_2(n)$ of winning a tournament with n players which is as great as, and in most cases greater than A's probability of winning a tournament with vector $(m_1, \ldots, m_k)$. This is achieved by considering two tournaments of type T_2, namely those whose vector m is of the form

$$T_2': \mathbf{m} = (x, y, 2^{t-1}, 2^{t-2}, \ldots, 1),$$

where $1 \leq x < K$ and $y = n - x - 2^t = K - x$, or

$$T_2'': \mathbf{m} = (x, y, 2^{t-2}, 2^{t-3}, \ldots, 1),$$

where $n/2 \geq x > K$ and $y = n - x - 2^{t-1}$.

Any tournament in T_2' is a first-stage descendant of T_2, and the probability of A winning any tournament in T_2'' is

$$\Pi_2' = p^t + \frac{p^{t-1}(p-1)}{n(n-x)}[(n+k-2x)2xp - (n-2x)(x-K)].$$

Straightforward calculation shows that $\Pi_2 > \Pi_2'$ for $n/2 > x > K$, and that $\Pi_2(n) = \Pi_2'(n)$ for $x = n/2$. An inductive argument completes the proof if one notes that any tournament with n players is a descendant of (is *dominated*! by) T_2 itself or of tournaments of the form T_2' or T_2''.□

We remark that we might have used the word domination in the sense of being a descendant, which is of course different from the idea of domination in the first three chapters. W. Maurer (1975) also uses the term dominance structure for tournaments, which is different from path or rank dominance. Maurer's results imply that $\Pi_2(n) \geq \Pi(n)$ for any tournament Π in a larger class than the simple model of 'one strong player, while other players are of equal strength.'

Since $\Pi + qE(R) = 1$ for any random tournament, Theorem 3A may be interpreted as follows: As $\Pi_4(n)$ is smaller for T_4 than for any other tournament, $E_4(R_4)$, the expected number of plays for the strongest player A, is maximized for T_4.

We conclude with a simple asymptotic result for repeated knock-outs using T_4. Let us suppose that after $n-1$ plays with T_4 the winner is obtained. We may continue the knock-out in a 'Markov' fashion, using the winner of the first replication to play randomly against one of the $n-1$ losers in the first replication. Calling such a tournament KOT, we see that by using every replication (rather than play) as the basic unit, we obtain a Markov chain by making the following definition.

DEFINITION B In playing KOT, let the system be said to be in *state* i in any replication ($i = 1, \ldots, R$) if the strong outlier A first entered that replication in game i.

Thus for the initial replication we have $a_1 = 2/n, a_2 = \ldots = a_{n-1} = 1/n$, for the probabilities a_i that the system is in state i. As the transition probabilities $\{p_{ij}\}$ from replication v to $v+1$ ($v \geq 1$) are (for $i = 1, \ldots, n-1$)

$$p_{i1} = p^{n-i} + \frac{1-p^{n-i}}{n-1}, \qquad p_{ij} = \frac{1-p^{n-i}}{n-1} \quad (j \geq 2),$$

it is immediate (cf. Feller 1968) that the stationary distribution $v = (v_1, \ldots, v_{n-1})$ is given by

$$[3a] \quad v_1 = \frac{1}{(n-2)q+1}, \qquad v_j = \frac{q}{(n-2)q+1} \quad (j \geq 2).$$

From [3a], which gives the asymptotic proportion of times the system is in state j ($j = 1, \ldots, n-1$), we can calculate the (asymptotic) proportion of games played by A in the tournament T_4. Given the system is in state j, and x is the random variable representing the number of games played by A in this replication, we have

$$P(x = i \mid j) = qp^{i-1} \qquad (i = 1, \ldots, n-j-1),$$

$$P(x = n-j \mid j) = p^{n-j-1}.$$

Thus, given the system is in state j,

$$[3b] \quad E(x) = (1-p^{n-j})/q.$$

Using [3a] and [3b] for states $j = 1, \ldots, n-1$, we see that [3a] also gives the proportions of games played by A and each of the other players in KOT. Of course [3b] with $j = 1$ is the maximum possible expected number of games played by A, and this would be achieved by making the player with highest score play the winner of each replication. Such a playing rule with emphasis on the highest score is denoted by QKOT. Indeed, the best rule we have found for selection procedures to pick the best player is a modification of QKOT with balance: At the end of replication i, the last winner either (a) has the highest score or (b) has not the highest score. In case (a) the last winner continues to play with balance, i.e. every player meets a randomly chosen player from among those he has played least, preference among those being given to the player who has played least in total. (A random choice is made if no such preference exists even in total plays.) In case (b) the last winner plays against one of the players with the highest score whom he has played the least, the same preferences (or lacking these the same random choice) as in (a) above being used. Thus, in both cases (a) and (b) the emphasis is on the player with the highest score starting a replication, and our numerical calculations and simulations show that such a knock-out (called BKOT for brevity), while at least partially balanced, retains the good features of KOT. We have found that slight modifications of BKOT are more effective than round-robins or pure knock-outs (in extensive simulations) for small, moderate, or large samples. For further details on this point, see Narayana and Hill (1974).

EXERCISES

1 Let $T(n, k)$ be the number of tournament vectors $(m_1, \ldots, m_k)$ with $n = 2^t + K$ players, where $0 \leq K < 2^t$.
(a) Verify the short table of values presented below, using for $k \geq t+1$

$$T(n, k) = T(n-1, k-1) + T(n-2, k-1) + \ldots + T([\tfrac{1}{2}(n+1)], k-1).$$

Values of $T(n, k)$ for $n, k \leq 9$

n \ k	2	3	4	5	6	7	8	9
3	1							
4	1	1						
5		2	1					
6		2	3	1				
7		1	5	4	1			
8		1	6	9	5	1		
9			6	15	14	6	1	
10			6	21	29	20	7	1

(b) Prove that $T(n, k) \leq \binom{n-3}{k-2}$ for all $k \geq 2$, $n \geq 3$.

(c) If $T_n = \sum_{k=1}^{n-1} T(n, k)$, prove that for $n \geq 11$

$$\frac{160}{256} < \frac{T_n}{2^{n-3}} \leq \frac{165}{256}.$$

2 Give an inductive proof that for all random tournaments (i.e. $p = \frac{1}{2}$) the probability that A and B meet in some round of the tournament is $2/n$, where n is the number of equally strong players.

3 Give direct combinatorial proofs of all the explicit values or recurrences for R_n^i and P_n^i for the tournaments T_1, T_2, T_3, T_4.

† 4 (Unsolved) For any tournament $(m_1, \ldots, m_k)$, let us consider Q_n^i, where Q_n^i is the probability that A meets B_1 or B_2 or ... B_i during the course of the tournament. Clearly when $p = \frac{1}{2}$, i.e. for a random knock-out tournament with all players of equal strength,

$$Q_n^1 = 2/n \quad \text{and} \quad Q_n^{n-1} = 1.$$

Further from the definition of Q_n^i

$$Q_n^i < Q_n^{i+1} \qquad (i = 1, \ldots, n-2).$$

Given any real number x in $(0, 1)$, and any tournament vector $(m_1, \ldots, m_k)$ of n players, clearly there exists a unique integer, $i = i(x; m_1, \ldots, m_k)$ say, such that

$$Q_n^{i-1} < x \leq Q_n^i \qquad (i = 1, \ldots, n-1).$$

Surely it is natural to define $Q_n{}^0 = 0$. When $p = \frac{1}{2}$, and $x = \frac{1}{2}$ so that $i(\frac{1}{2}; m_1, \ldots, m_k)$ corresponds to the 'median,' we have found, by computer for all $n \leq 256$, the remarkable result that

$$i(\tfrac{1}{2}; m_1, \ldots, m_k) = \{(n-1)/3\},$$

where $\{(n-1)/3\}$ represents the smallest integer $\geq (n-1)/3$. How about the more general conjecture

$i(1/2^l;\, m_1, \ldots, m_k) = \{(n-1)/(2^{l+1}-1)\}$ independently of $m_1, \ldots, m_k$?

5 If $\Pi_1{}^*\,(m_1, \ldots, m_k)$ is the probability that a strong player wins the tournament with vector $(m_1, \ldots, m_k)$ when he is the only strong player present, and $\Pi_2{}^{**}\,(m_1, \ldots, m_k)$ is the probability that a strong player wins the tournament with vector $(m_1, \ldots, m_k)$ when there are two strong players present, show that

$$\Pi_2{}^{**}(m_1, \ldots, m_k) = \left\{\binom{n-2}{2m_1-2}\Big/\binom{n}{2m_1}\right\} \cdot \left\{\frac{1}{2m_1-1} \cdot \frac{1}{2} \cdot \Pi_1{}^*(m_2, \ldots, m_k) + \frac{2m_1-2}{2m_1-1} p[p\Pi_2{}^{**}(m_2, \ldots, m_k) + q\Pi_1{}^*(m_2, \ldots, m_k)]\right\} + \left\{2\binom{n-2}{2m_1-1}\Big/\binom{n}{2m_1}\right\} \cdot \{p\Pi_2{}^{**}(m_2, \ldots, m_k)\}$$

$$+\left\{\binom{n-2}{2m_1-1}\Big/\binom{n}{2m_1}\right\}\cdot q\Pi_1{}^*(m_2,\ldots,m_k)$$

$$+\left\{\binom{n-2}{2m_1}\Big/\binom{n}{2m_1}\right\}\cdot \Pi_2{}^{**}(m_2,\ldots,m_k).$$

6 Interchanging the role of winner and loser in KOT, and denoting this tournament by TOK, show that the proportion of replications initiated by the strong outlier in a tournament consisting of a number of KOT replications followed by an equal number of TOK replications approaches asymptotically a quantity greater than $2/n$. (This is the proportion of plays in a round robin for all players.) Tournaments QTOK and BTOK are similarly defined relative to QKOT and BKOT.

†7 Consider a tournament with i strong players and $2n-i$ weak players, paired off at random for round 1. Each strong player always beats a weak player. Denoting by $P(2n; i, k)$ the probability that exactly $2k$ strong players are paired off in round 1, show that

$$P(2n; i, k) = \frac{i_{(2k)}n_{(k)}}{2^{2k-i}k!(2n)_{(i)}} \qquad ([\tfrac{1}{2}i] \geq k \geq 0).$$

Note that the above expression yields the probability that exactly k strong players lose in round 1, i.e. letting $i-k=j$, the transition probability that exactly j strong players survive round 1 is

$$(1)\quad t(2n; i, j) = \begin{cases} \dfrac{i_{(2i-2j)}2^{2j-i}n_{(i)}}{(i-j)!(n-j)_{(i-j)}(2n)_{(i)}} & (2j \geq i \geq j), \\ 0 & \text{otherwise.} \end{cases}$$

†8 Specialize (1) to the classical case with 2^n players, denoting by $t_{ij}^{(n)}$ the transition probability $P(2^n; i, j)$. If $T(n) = \{t_{ij}^{(n)}\}$ is the $2^n \times 2^{n-1}$ transition matrix, and $C(n) = \{C_{ij}^n\}$ is the $2^n \times n$ matrix with entries

$$C_{ij}^n = 2^{n-j+1} - \frac{1}{2^{j-1}}\frac{(2^n - 2^{j-1})_{(i)}}{(2^n-1)_{i-1}},$$

prove that $T(n+1)C(n) = C'(n+1)$ where $C'(n+1)$ is $C(n+1)$ with its first column omitted.

REFERENCES

1 CAPELL, P., and T. V. NARAYANA. 1970. 'On knock-out tournaments,' *Can. Math. Bull. 13*, 105–9

2 DAVID, H. A. 1963. *The Method of Paired Comparisons*. Charles Griffin and Company Ltd., London

3 FELLER, W. 1968. *An Introduction to Probability Theory and Its Applications*, Vol. I. John Wiley and Sons, New York

4 HOPKINS, M. J. D. 1969. 'A comparison of tournaments.' Unpublished M.Sc. thesis, University of Alberta

5 MAURER, W. 1975. 'On most effective tournament plans with fewer games than competitors,' *Ann. Statist. 3*, 717–27

6 MOON, J. W. 1968. *Topics on Tournaments*. Holt, Rinehart and Winston, New York

7 NARAYANA, T. V. 1968. 'Quelques résultats relatifs aux tournois "knock-out" et leurs applications aux comparaisons aux paires,' *Comp. Rend. Acad. Sci. Paris 267*, 32–3

8 NARAYANA, T. V., and J. HILL. 1974. 'Contributions to the theory of tournaments III,' *Proc. Fifth Nat. Math. Conf., Shiraz, Iran*, 187–221

9 NARAYANA, T. V., and J. ZIDEK. 1969. 'Contributions to the theory of tournaments I,' *Cahiers du Bur. Univ. de Rech. Opér. 13*, 1–18

10 TUCKER, H. G. 1962. *Introduction to Probability and Mathematical Statistics*. Academic Press, New York

V

A miscellany of further research problems*

As ocean sweeps the laboured mole away;
While self-dependent power can time defy,
As rocks resist the billows and the sky.

Goldsmith, *The Deserted Village* ('touched' by Coleridge?)

We conclude the volume with a discussion of a variety of problems suggested by recent developments in dominance theory. Section 1 deals with selection problems in paired comparisons, where a simple application of Feller's 'ladder depth' idea gives a rich class of tournament designs involving knock-outs. A variation of BKOT yields extremely efficient selection procedures when tournaments are viewed as random walks in n-dimensions. We return in the remaining sections to problems involving Young chains and non-parametric tests, as the full power of dominance methods has yet to be utilized in statistics, whether in non-parametric testing or in sequential procedures involving dominance in the continuous case. The exercise indicates how the *théorème d'équerre* of Frame, Robinson, and Thrall (1954) defines the basic (or irreducible?) non-parametric tests, which are of great use to statisticians in Bayesian non-parametrics.

1
A COMPARISON OF SELECTION PROCEDURES

1.1. *The design of knock-out tournaments*

We first state a few theoretical results from statistics, giving references to the interested reader for details.

Consider repeated knock-outs played according to any fair design D, in the sense that if all players are of equal strength, then the (random) number of trials in which players $1, \ldots, n$ participate – say $N_1, \ldots, N_n$ – prove to be exchangeable random variables. Then clearly N_{Si} ($i = 1, \ldots, n$), the random number of trials in

* The reader should be familiar with reference 1 at the end of this chapter to understand the applications.

which player i wins among his N_i plays, is also an exchangeable variable. Suppose the tournament with design D has been independently repeated m times and let $N_i^{(v)}$, $N_{Si}^{(v)}$ denote the independent copies of N_i, N_{Si} respectively $(i = 1, \ldots, n)$ determined by the vth replication. Let T_{im} and S_{im} $(i = 1, \ldots, n; m \geq 1)$ be defined by

$$[1a] \quad T_{im} = \sum_{v=1}^{m} N_i^{(v)}, \qquad S_{im} = \sum_{v=1}^{m} N_{Si}^{(v)}.$$

LEMMA 1A (i) *For the one strong outlier model we consider,*

$$(T_{1m}, S_{1m}, \ldots, T_{nm}, S_{nm})$$

is a sufficient statistic.

(ii) *When* $m \to \infty$,

$$\left(\frac{[T_{1m} - 2S_{1m}]^+}{\sqrt{T_{1m}}}, \ldots, \frac{[T_{nm} - 2S_{nm}]^+}{\sqrt{T_{nm}}}\right) \overset{L}{\sim} \alpha(Z_1{}^+, \ldots, Z_n{}^+)$$

where $\alpha(Z) = N(0, \Sigma)$, *i.e. the normal with mean* 0 *and covariance* Σ. *Here* $\Sigma = (\Sigma_{ij})$ *where*

$$\Sigma_{ij} = \begin{cases} 1 & \text{if } i = j, \\ -\dfrac{1}{n-1} & \text{if } i \neq j; \end{cases}$$

$\overset{L}{\sim}$ *denotes asymptotic equality in law; and* $w^+ = \max(w, 0)$.

□For the proof, see Narayana and Zidek (1969).□

Following results of Zidek (unpublished), the class of all Bayes invariant tests, and their asymptotic behaviour as well as subset selection procedures for our model, are given in Theorems 1, 2, 3 of Narayana and Zidek (1969).

David (1963) provides a table (his Appendix Table 5) for selecting the best subset of players when n replications of a round-robin (RR) with t players are made. This table gives the values of v for various n, t and a given probability or confidence P^* (with $P^* = 0.75, 0.90, 0.95, 0.99$). Here P^* is the minimum probability with which one can be certain that the best player is in the selected subset, where the 'best' subset includes all players with score S_{jm} where

$$[1b] \quad \max_j S_{jm} - v \leq S_{jm}.$$

David (1963, p. 113) also gives tables for the same confidences for the smallest number of round-robin replications required to ensure that the selection of the player with maximum score actually yields the best player with at least the predetermined confidence for our model. For economy in our comparisons of various procedures, we shall use David's total number of plays from his Table 4 and determine from the same simulations the average expected best subset size using, from David's Table 5, the appropriate v for both round-robins and various knock-outs such as KOT and QKOT.

1.2. *Ladder depths for tournament design*

It is convenient in describing knock-out procedures to score $+1$ for the winner of any play and -1 for the loser, so that at every stage the sum of the scores of all players equals zero. We thus obtain a random walk in $(n-1)$-dimensions defined by the score vector $(S_1, \ldots, S_n)$ of the n players, where $\sum S_i = 0$. At each game played by the outlier, his score increases by one with probability p and decreases by one with probability $q = 1-p$. Thus his score describes a simple random walk with absorbing barrier at $-k$. It can be shown (Feller 1968) that the probability of the outlier reaching a 'depth' $-k$ is $\leq (q/p)^k$. Since there are as many as $n-1$ eliminations in an n player tournament, we choose k as the smallest integer such that

$$\{1-(q/p)^k\}^{n-1} \geq P$$

for a prespecified probability P. Table 8 gives values of k for various cases. These verify the values first found by Brett at the University of Western Australia (personal communication, 1971).

Suppose it is desired to eliminate from competition any player whose score drops to a prespecified value $-k$. This process is to be repeated until $n-1$ players have been eliminated. The survivor is to be called the winner. We can also consider three possible courses of action to be followed after a player is eliminated:

A. Upon the elimination of one of n players, the scores of the remainder are returned to zero and a new tournament is begun among the $n-1$ survivors. This process is repeated until a single player is left. The probability of the outlier being the first to reach $-k$ is obviously less than $(q/p)^k$. This relation holds for all $n-1$ eliminations, so the probability of the outlier winning is greater than

$$\{1-(q/p)^k\}^{n-1}.$$

A given level of probability P can thus be attained by appropriate choice of k (see Table 8).

B. Upon elimination, the scores made against the eliminated player by the surviving players are removed from the survivors' scores. The survivors retain, however, the scores made among themselves, and continue to play until only the winner remains. Suppose the original scores of the survivors are $(S_1, S_2, \ldots, S_{n-1})$ and the corrected scores $(S_1', S_2', \ldots, S_{n-1}')$; note that $\sum S_i = k$, but $\sum S_i' = 0$. If the first player is the strong outlier, then $E(S_1) > 0$ and hence the probability of the

TABLE 8

Smallest k such that $\{1-(q/p)^k\}^{n-1} \geq C$

C \ n	2	3	4	5	6	7	8	9	10
$p = 0.55$									
0.75	7	11	12	14	15	16	17	17	18
0.90	12	15	17	19	20	21	21	22	23
0.95	15	19	21	22	23	24	25	26	26
0.99	23	27	29	30	31	32	33	34	34
$p = 0.65$									
0.75	3	4	4	5	5	5	6	6	6
0.90	4	5	6	6	7	7	7	8	8
0.95	5	6	7	8	8	8	8	9	9
0.99	8	9	10	10	11	11	11	11	11
$p = 0.75$									
0.75	2	2	3	3	3	3	3	4	4
0.90	3	3	4	4	4	4	4	4	5
0.95	3	4	4	4	5	5	5	5	5
0.99	5	5	6	6	6	6	6	7	7

strong player descending to $-k$ should be less than $(q/p)^k$, provided he is not already eliminated. Thus it is reasonable to claim that method B will enable the outlier to win with a higher probability than method A.

C. All scores are retained, and corrections made simply by the relation $S_i' = S_i - k/(n-1)$, so that once again $\sum S_i' = 0$. Clearly, method C exhibits the same advantages as mentioned for B. In addition the outlier presumably will have won against the eliminated player with a higher probability than the other survivors, and hence method C can be expected to work slightly better than B.

We conclude our discussion of ladder depths for tournament design by presenting a table (Table 9) comparing RR, KOT, QKOT, and BKOT*. BKOT* is BKOT with one slight modification as follows: Until the first elimination the order of play is reversed, that is, the loser plays through and a player of least score starts each replication. The purpose of this measure using 'BTOK' for one elimination is to dispose quickly of a totally worthless player before starting BKOT proper. In practice, at the first elimination a test may be made regarding the equality of the scores (see, for example, David 1963, Narayana and Zidek

TABLE 9

Proportion of wins and best sample size in 100 or more simulations for round-robin, knock-out, and eliminatory knock-out with percentage savings ($p = 0.75$ = 'confidence')

n	RR		KOT		BKOT *†		% savings
(a) $p = 0.55$							
4	0.75	2.30	0.745	2.21	0.77	1.17	36.47
5	0.77	2.76	0.83	2.71	0.85	1.30	27.35
6	0.748	3.60	0.81	3.07	0.87	1.30	28.39
7	0.745	3.94	0.695	3.53	0.87	1.53	18.95
8	0.70	4.05	0.74	3.97	0.88	1.56	18.06
(b) $p = 0.65$							
4	0.74	2.70	0.755	2.29	0.85	1.27	29.69
5	0.80	2.92	0.80	2.47	0.87	1.45	20.58
6	0.74	3.27	0.808	2.33	0.87	1.37	23.29
7	0.722	3.72	0.895	3.04	0.91	1.36	24.65
8	0.782	4.52	0.835	3.25	0.86	1.99	16.27
(c) $p = 0.75$							
4	0.807	2.53	0.77	2.29	0.89	1.51	21.44
5	0.793	3.39	0.89	2.31	0.89	1.75	17.13
6	0.788	3.16	0.91	2.3	0.90	1.37	27.80
7	0.755	4.43	0.89	2.61	0.96	1.13	39.10
8	0.905	3.67	0.95	1.89	0.95	1.13	35.56

† 150 simulations.

NOTE: QKOT is given in Table 10.

1969), the result of this test governing subsequent action. In Table 9, we used one BTOK for the first elimination and BKOT for all subsequent eliminations, *method C of adjusting the scores being employed at every elimination.* Finally, for BKOT* the non-eliminated players are all included in the best subset, so that the probability of correct selection of the best player is a lower bound for the best player to be in the selected subset.

It is fairly evident from Table 9 that the 'stochastic approximation' type BKOT* is a near optimal design for all selection problems. In practice the following variations permit a large class of designs to be at the disposal of a statistician: (i) variations in numbers of BTOK used, based on a statistical test; (ii) variations in the method of adjusting scores, such as giving only half-weight to an eliminated player's score and eliminating even this effect at the next elimination; (iii) use of the substantial saving to locate a possibly previously eliminated best player. For more detail on these points, see Narayana and Hill (1974).

TABLE 10

Comparison of KOT and QKOT with number of wins (out of 100), means, and variances of selected subset size ($P = 0.75$ = 'confidence')

n	KOT			QKOT		
(a) $p = 0.55$						
4	74.5	2.21	0.93	77	2.24	1.22
5	83	2.71	2.39	85†	2.38	1.48
6	81	3.07	2.69	82†	2.52	2.49
7	69.5	3.53	3.55	79†	3.34	3.78
8	74	3.97	4.83	70†	3.88	4.07
(b) $p = 0.65$						
4	75.5	2.29	1.43	78	2.01	1.39
5	80	2.47	2.17	77	2.32	1.86
6	80.83	2.33	2.50	84	2.55	2.81
7	89.5	3.04	3.76	85.5	2.78	4.09
8	83.5	3.25	5.33	77	3.52	5.43
(c) $p = 0.75$						
4	77	2.29	1.39	77	1.97	1.37
5	89	2.31	2.15	81.5	2.71	2.53
6	91	2.3	2.87	84.5	2.29	3.23
7	89	2.61	3.60	87	2.71	4.75
8	95	1.89	2.48	92	2.05	3.05

† Based on 50 simulations only.
NOTE The values for KOT from Table 9 are included for easy comparison.

A note of warning is added to those who glibly recommend 'asymptotically efficient' procedures without examining their performance for small or even moderately large samples. Truncated procedures using ladder depth with KOT and QKOT perform extremely poorly – in spite of the asymptotic equivalence of QKOT and BKOT. Extensive simulation leaves no doubt on this point, and the safeguard of balance in BKOT cannot be neglected with impunity.

2

THE NUMBERS $\binom{n}{r}\binom{n}{r-1}\Big/n$

The numbers (n, r) given by

$$[2a] \quad (n, r) = \binom{n}{r}\binom{n}{r-1}\Big/n \qquad (r = 1, \ldots, n),$$

which we have encountered in [2i], Chapter II, were first introduced in the context of domination by Narayana (1955). They represent the number of paths from $(0,0)$ to (n,n) lying entirely below the diagonal and having $r-1$ (r) turns of the type vertical followed by horizontal (horizontal followed by vertical) step. They also evaluate the number of Young chains with $r-1$ switchbacks from $(0,0)$ to (n,n). From well-known properties of the Catalan numbers,

$$[2b] \quad \sum_{r=1}^{n} (n,r) = \frac{1}{n+1}\binom{2n}{n},$$

where the right-hand side is the number of ways of filling the Young tableau of 2 rows and n columns. From the Frame-Robinson-Thrall *théorème d'équerre* the right-hand side is evaluated as

$$2n!/(n+1)!n!$$

and Kreweras (1967) first noted the Young chain interpretation of [2b]. Indeed Narayana (1959) essentially proves, for $a = (n,n)$, $b = (0,0)$, that

$$[2c] \quad |K_r(a,b)| = \frac{1}{n+r+1}\binom{n+r+1}{r}\binom{n+r+1}{r+1} \qquad (r = 0, 1, 2, \ldots,);$$

and Kreweras (1967) pointed out the remarkable identity

$$[2d] \quad \frac{\binom{n+r+1}{r}\binom{n+r+1}{n}}{n+r+1} = \sum_{s=0}^{r} \frac{\binom{n}{r-s}\binom{n}{r-s+1}}{n}\binom{2n+s}{2n}$$

corresponding to Kreweras's Theorem. For further interpretations of the numbers (n,r) we refer to the bibliography at the end of the book, specifically references 5, 6, and 16. One unsolved problem connected with (n,r) is the distribution time of the first, second, or nth turn or switchback. This problem has been touched upon by Kreweras (exercise 11, Chapter I), who also considers generalizations with diagonal steps possible.

3
WEAK INADMISSIBILITY OF TESTS

From the combinatorial developments arising through dominance, it is clear that to every Young chain corresponds an irreducible or basic non-parametric test. Conversely to unions of Young chains (or dominance chains more generally) correspond dominance tests. We shall illustrate these concepts, as well as weak inadmissibility of tests, by an example.

EXAMPLE Let $m = 3$, $n = 2$ and consider the Lehmann distribution for paths for

$k=2$. The possible paths A with their probabilities (multiplied by 105) and ranks (S_1, S_2) are given below.

Path	Rank	Probability ($\times 10^5$)
(3, 3)	(1, 2)	3
(2, 3)	(1, 3)	4
(1, 3)	(1, 4)	5
(0, 3)	(1, 5)	6
(2, 2)	(2, 3)	8
(1, 2)	(2, 4)	10
(0, 2)	(2, 5)	12
(1, 1)	(3, 4)	15
(0, 1)	(3, 5)	18
(0, 0)	(4, 5)	24

The five Young chains can be easily listed, being defined in path notation by

$$Y_1:\ 00-01-02-03-13-23-33$$
$$Y_2:\ 00-01-02-12-13-23-33$$
$$Y_3:\ 00-01-02-12-22-23-33$$
$$Y_4:\ 00-01-11-12-13-23-33$$
$$Y_5:\ 00-01-11-12-22-23-33$$

Consider the dominance tests or unions of dominance chains

$$W = 00,\ 01,\ 02 \cup 11,\ 12 \cup 03,\ 22 \cup 13,\ 23,\ 33$$
$$T_1 = 00,\ 01,\ 11,\ 12,\ 22,\ 22 \cup 03,\ 22 \cup 13,\ 23,\ 33$$
$$T_2 = 00,\ 01,\ 11,\ 11 \cup 02,\ 12,\ 13,\ 22 \cup 13,\ 23,\ 33$$
$$M = 00,\ 01,\ 11,\ 11 \cup 02,\ 12,\ 22,\ 22 \cup 03,\ 22 \cup 13,\ 23,\ 33$$

M is the most powerful rank test using Y_1, Y_5 in its dominance structure as follows (take corresponding unions):

$$00-01-01-02-02-02-03-13-23-33,$$
$$00-01-11-11-12-22-22-22-23-33.$$

Similarly W is the union of all Young chains, being the Wilcoxon test, whereas T_1 and T_2 are dominance tests using Y_1, Y_3 and Y_2, Y_5 respectively. In the class of tests $\{M, T_1, T_2, W\}$ only M is admissible à la Wald; however, in the class $\{T_1, T_2, W\}$, W is weakly inadmissible since at every level attained by W(randomized or not) there is a better test among T_1, T_2 with actually higher power at some levels. Indeed, using the appropriate test of T_1, T_2 *taking into account the level* gives the same power as M. So (T_1, T_2) are an essentially weakly complete class among all tests not including M.

The author conjectures that: (i) for Lehmann alternatives, W and C_1, the Fisher-Yates-Terry-Hoeffding test, are inadmissible à la Wald; (ii) for normal shift alternatives W is inadmissible *weakly* in the class $W \cup C_1 \cup DR(a)$, where $DR(a)$ represents the dominance-ratio tests, $a \geq 0$ (see Chapter III, exercise 13).

EXERCISE

1 Show that in the case $m = n = 3$, letting $a = 333$ and $b = 000$, we have

$$|K_r(a,b)| = \frac{1}{30}\binom{r+5}{3}\binom{r+4}{4}\binom{r+3}{2} \qquad (r = 0, 1, 2, \ldots),$$

and the Young chains with r switchbacks satisfy

$$|Y_r(a,b)| = \binom{5}{r}\binom{5}{r+1}\Big/5 \qquad (r = 0, 1, \ldots, 4).$$

Verify that the number of Young chains is

$$|Y(a,b)| = \sum_{r=0}^{4} |Y_r(a,b)| = 42$$

directly from the Frame-Robinson-Thrall theorem. Show that the Wilcoxon test is weakly inadmissible in a suitable class (without constructing the most powerful test explicitly) against Lehmann alternatives with $k = 3$.

REFERENCES

1 David, H. A. 1963. *The Method of Paired Comparisons*. Charles Griffin and Co., London

2 Feller, W. 1968. *An Introduction to Probability Theory and Its Applications*, Vol. I. John Wiley and Sons, New York

3 Frame, J. S., G. de B. Robinson, and R. M. Thrall. 1954. 'The hook graphs of the symmetric groups,' *Can. J. Math. 6*, 316–23

4 Kreweras, G. 1967. 'Traitement simultané du "Problème de Young" et du "Problème de Simon Newcomb,' *Cahiers du Bur. Univ. de Rech. Opér. 10*, 23–31

5 Narayana, T. V. 1955. 'Sur les treillis formés par les partitions d'un entier; leurs applications à la théorie des probabilités,' *Comp. Rend. Acad. Sci. Paris 240*, 1188–9

6 – 1959. 'A partial order and its application to probability theory,' *Sankhya 21*, 91–8

7 Narayana, T. V., and J. Hill. 1974. 'Contributions to the theory of tournaments III,' *Proc. Fifth Nat. Math. Conf., Shiraz*, 187–221

8 Narayana, T. V., and J. Zidek. 1969. 'Contributions to the theory of tournaments II,' *Rev. Roum. Math. Pures et Appl. 10*, 1563–76

9 Zidek, J. 'A representation of Bayes invariant procedures in terms of Haar measure' (unpublished)

APPENDIX

On some convolution identities from lattice path combinatorics*

L'homme n'est qu'une faible lueur dans la tempête,
mais cette lueur résiste et cette lueur est tout.

Henri Poincaré

In this appendix, convolution identities which arise from path combinatorics and thus may conveniently be termed 'identities from path combinatorics' are presented. Essentially the appendix represents a compilation of earlier work (more or less inspired by dominance) of Drs. Handa and Mohanty. Two identities related to trees are also given.

1

The classical Euler summation formula or Vandermonde's convolution formula states that

$$[1a]\quad \sum_{k=0}^{n}\binom{a}{k}\binom{b}{n-k}=\binom{a+b}{n}$$

for any a and b. When a and b are non-negative integers, identity [1a] has a simple and elegant interpretation in terms of minimal lattice paths (hereafter called paths). The right-hand side represents the number of paths from the origin to $(a+b-n, n)$. Since each such path has to pass through $(a-k, k)$ for some $k, k = 0, 1, \dots, n$, we get the expression on the left-hand side. It is known (see references 2, 10, 23, 27) that the number of paths from the origin to the point $(a+(r-1)n, n)$ not touching the line $x=(r-1)y$ (r being a positive integer) except at the origin is

$$\frac{a}{a+rn}\binom{a+rn}{n}.$$

Using the above idea of convolution, one can easily write the identity

$$[1b]\quad \sum_{k=0}^{n}\frac{a}{a+rk}\binom{a+rk}{k}\frac{b}{b+r(n-k)}\binom{b+r(n-k)}{n-k}=\frac{a+b}{a+b+rn}\binom{a+b+rn}{n},$$

* This appendix is based on a paper presented by S. G. Mohanty and B. R. Handa at the 80th Summer Meeting of the American Mathematical Society, 24–27 August 1976.

which is seen to be true even for general a, b, and r (see reference 5) and is a generalization of Vandermonde's convolution. The proof of the general case uses the usual generating function method. Gould has developed other similar identities in references 6, 7, 8.

From the path point of view, formula [1b] has been generalized further. A path from the origin to (m, n) can be represented (see Chapter I, [1a]) by a vector $(x_0, x_1, ..., x_n)$ satisfying $0 = x_0 \leq x_1 \leq ... \leq x_n$ where $x_i\,(i = 0, 1, ..., n)$ is the distance of the path from $(m, n-i)$. Thus the paths which lie below and do not cross the path represented by $(a_0, a_1, ..., a_n)$, $0 = a_0 \leq a_1 \leq ... \leq a_n$, are equivalent to the set of vectors $(x_1, ..., x_n)$ dominated by $(a_1, ..., a_n)$. The number of such vectors is

$$[1c] \quad \sum_{x_1=0}^{a_1} \cdots \sum_{x_{n-1}=x_{n-2}}^{a_{n-1}} \sum_{x_n=x_{n-1}}^{a_n} 1;$$

this in turn equals (either directly or by Kreweras's Theorem 1A, Chapter II)

$$[1d] \quad \det_{n \times n} \|c_{ij}\| \qquad \text{where } c_{ij} = \binom{a_i+1}{j-i+1}.$$

Interestingly, if we replace the summation signs by integral signs in [1c], we get another determinant

$$[1e] \quad \det_{n \times n} \|d_{ij}\| \qquad \text{where } d_{ij} = \frac{a_i^{j-i+1}}{(j-i+1)!}.$$

Here the a_i's need not be integers. Similar to [1b], we get the identity

$$[1f] \quad \sum_{k=0}^{n} \frac{a}{a+rk}\frac{(a+rk)^k}{k!}\frac{b}{b+r(n-k)}\frac{(b+r(n-k))^{n-k}}{(n-k)!} = \frac{a+b}{a+b+rn}\frac{(a+b+rn)^n}{n!}.$$

Identity [1f] is known to be an Abel-type convolution formula. The analogous identity corresponding to [1a] is

$$[1g] \quad \sum_{k=0}^{n} \frac{a^k}{k!}\frac{b^{n-k}}{(n-k)!} = \frac{(a+b)^n}{n!},$$

which is trivially the well-known formula for binomial expansion.

In references 6 and 7 Gould gives a unified treatment by introducing 'C-coefficients' of which

$$\frac{a}{a+rn}\binom{a+rn}{n} \quad \text{and} \quad \frac{a}{a+rn}\frac{(a+rn)^n}{n!},$$

as denoted by $A_n(a, r)$ and $B_n(a, r)$, are special cases. By our path analysis, the above synthesis seems logical and natural. Gould has defined C-coefficients which satisfy

$$\sum_{n=0}^{\infty} C_n(\alpha, \beta)z^n = x^\alpha,$$

$$\sum_{n=0}^{\infty} G_n(\alpha, \beta)z^n = x^\alpha g(x, \beta),$$

where

$$z = h(x)/x^\beta$$

and

$$G_n(\alpha, \beta) = \frac{\alpha+n\beta}{\alpha} C_n(\alpha, \beta).$$

His two other identities of interest are, from reference 8,

$$[1h] \quad \sum_{k=0}^{n} (-1)^k A_k(c, tb)A_{n-k}(a+bk-k, (1-t)b) = A_n(a-c, (1-t)b),$$

and

$$[1i] \quad \sum_{k=0}^{n} (-1)^k B_k(c, tb)B_{n-k}(a+bk, (1-t)b) = B_n(a-c, (1-t)b),$$

which are responsible for the derivation of certain inverse series relations. We remark that identity [1h] is a path-type convolution, though it does not look so.

Using path interpretation, it is obvious that one can form convolution identities from [1d] or [1e]. The question whether such identities would be true in general has been answered in reference 21. Following the idea of C-coefficients, we define g-coefficients as follows. Let

$$\delta_n^i = \begin{cases} 1 & \text{for } n = i, \\ 0 & \text{for } n \neq i. \end{cases}$$

Consider a sequence $\{f_n(\xi)\}$ with property

$$f_n(\xi) = 1 \quad \text{and} \quad f_n(0) = \delta_n^0.$$

Letting $\phi(s, \xi)$ represent the generating function of the sequence, we say that ϕ is additive with respect to ξ if $\phi(s, \xi+\alpha) = \phi(s, \xi)\phi(s, \alpha)$. For any sequence $b_1, b_2, \ldots,$ a g-coefficient denoted by $g(0, b_i, \ldots, b_n)$, $i = 1, \ldots, n$, $n = 1, 2, \ldots,$ is the $(n-i+1)$th order determinant

$$[1j] \quad \begin{vmatrix} f_1(b_n) & f_2(b_{n-1}) & \cdots & f_{n-i+1}(b_i) \\ f_0(b_n) & f_1(b_{n-1}) & \cdots & f_{n-i}(b_i) \\ 0 & f_0(b_{n-1}) & \cdots & f_{n-i-1}(b_i) \\ \vdots & \vdots & & \vdots \\ 0 & 0 & \cdots & f_1(b_i) \end{vmatrix}$$

and $g(0) = 1$. Note that $g(0, b_i, \ldots, b_n) = 0$ if $b_i = 0$. It has been proved that a necessary and sufficient condition for the identities

$$[1k] \quad \begin{cases} g(0, a_i - b_i, a_{i+1} - b_i, \ldots, a_n - b_i) \\ \qquad + \sum_{j=i}^{n-1} g(0, a_{j+1} - b_{j+1}, \ldots, a_n - b_{j+1}) g(0, b_i, \ldots, b_j) \\ \qquad + g(0, b_i, \ldots, b_n) = g(0, a_i, \ldots, a_n) \\ \qquad\qquad\qquad (i = 0, \ldots, n-1; n = 2, 3, \ldots), \\ g(0, a_n - b_n) + g(0, b_n) = g(0, a_n) \qquad (n = 1, 2, \ldots), \end{cases}$$

to hold for any a_i's and b_i's is that ϕ is additive with respect to ξ.

Convolution identities involving [1d] or [1e] are established when in particular $f_n(\xi) = \binom{\xi}{n}$ or $f_n(\xi) = \xi^n/n!$ respectively. It is seen that Gould's identities become special cases of [1k]. For example, when $a_r = b(t-1)(r-1) + c - a$ and $b_r = c + (r-1)(tb-1)$ for all r, [1k] reduces to [1h]. An important observation is that the condition ϕ is additive is equivalent to [1a] when $f_n(\xi) = \binom{\xi}{n}$ and to [1g] when $f_n(\xi) = \xi^n/n!$ Thus, very surprisingly, [1a] or [1g] seems to be the fundamental identity. Some new identities have been derived in reference 21, as special cases of [1k].

Another direction of extension is to think of paths in higher dimension. The number of paths from the origin to the point $\left(a + \sum_{i=1}^{k} (r_i - 1)n_i, n_1, \ldots, n_k\right)$ in $(k+1)$-dimensional space not touching the hyperplane $x_0 = \sum_{i=1}^{k} (r_i - 1)x_i$ except at the origin is known to be (see reference 19)

$$[1l] \quad \frac{a}{a + \sum_{i=1}^{k} r_i n_i} \begin{bmatrix} a + \sum_{i=1}^{k} r_i n_i \\ n_1, \ldots, n_k \end{bmatrix}$$

Here

$$\begin{bmatrix} a + \sum_{i=1}^{k} r_i n_i \\ n_1, \ldots, n_k \end{bmatrix} = \frac{\left(a + \sum_{i=1}^{k} r_i n_i\right)\left(a + \sum_{i=1}^{k} r_i n_i - 1\right) \ldots \left(a + \sum_{i=1}^{k} (r_i - 1)n_i + 1\right)}{\prod_{i=1}^{k} n_i!}.$$

Thus an immediate extension of [1b] is seen to be

$$[1m]\quad \sum_{j_k=0}^{n_k}\cdots\sum_{j_1=0}^{n_1}\frac{a}{a+\sum_{i=1}^{k}r_ij_i}\begin{bmatrix}a+\sum_{i=1}^{k}r_ij_i\\ j_1,\ldots,j_k\end{bmatrix}$$

$$\times\frac{b}{b+\sum_{i=1}^{k}r_i(n_i-j_i)}\begin{bmatrix}b+\sum_{i=1}^{k}r_i(n_i-j_i)\\ n_1-j_1,\ldots,n_k-j_k\end{bmatrix}$$

$$=\frac{a+b}{a+b+\sum_{i=1}^{k}r_in_i}\begin{bmatrix}a+b+\sum_{i=1}^{k}r_in_i\\ n_1,\ldots,n_k\end{bmatrix}$$

Similarly,

$$[1n]\quad \sum_{j_k=0}^{n_k}\cdots\sum_{j_1=0}^{n_1}\frac{a}{a+\sum_{i=1}^{k}r_ij_i}\frac{\left(a+\sum_{i=1}^{k}r_ij_i\right)^a}{\prod_{i=1}^{k}j_i!}$$

$$\times\frac{b}{b+\sum_{i=1}^{k}r_i(n_i-j_i)}\frac{\left(b+\sum_{i=1}^{k}r_i(n_i-j_i)\right)^b}{\prod_{i=1}^{k}(n_i-j_i)!}=\frac{a+b}{a+b+\sum_{i=1}^{k}r_in_i}\frac{\left(a+b+\sum_{i=1}^{k}r_in_i\right)^{a+b}}{\prod_{i=1}^{k}n_i!},$$

where

$$a=\sum_{i=1}^{k}j_i,\qquad b=\sum_{i=1}^{k}(n_i-j_i).$$

Following Gould's method, one can establish [1m] and [1n] for any $a, b, r_1, \ldots, r_k$ as in reference 15. Higher-dimensional extensions of [1h] or [1i] and of C-coefficients also exist (see reference 20).

Determinants [1d] and [1e] can be generalized further, as indicated essentially by Narayana in reference 23. Consider the matrix $(x_{ij})_{n\times m}$ such that

$$0\le x_{il}\le\ldots\le x_{im}\le a_i\qquad (i=1,\ldots,n),$$

and

$$x_{ij}\le x_{i+1,j}\qquad (i=1,\ldots,n-1;j=1,\ldots,m).$$

The number of such matrices is given by the determinant

$$[1o]\quad \det_{n\times n}\|c_{ij}^{(m)}\|\qquad\text{where } c_{ij}^{(m)}=\begin{bmatrix}a_i+m\\ m+j-i\end{bmatrix}.$$

A similar generalization of $\|d_{ij}\|$ is also possible but is omitted. The answer to the question whether we can form the convolution identity for $m>1$ is negative, as is seen from the following example for $n=1$, $m=2$:

$$\begin{bmatrix} a_1+2 \\ 2 \end{bmatrix} \neq \begin{bmatrix} b_1+2 \\ 2 \end{bmatrix} + \begin{bmatrix} a_1-b_1+2 \\ 2 \end{bmatrix} \text{ in general.}$$

Finally it may be worth while to note that we have seen a class of identities related to lattice paths which are valid even beyond the interpretation of paths. However, the notion of paths has helped to create identities which in this sense may be justifiably called identities from path combinatorics.

2

In this section we record a few orthogonal relations and inversion formulae arising out of them. An excellent reference on inverse relations is Riordan's book (reference 25). From [1j], we easily obtain the following orthogonal relations:

[2a] $$\sum_{j=i+1}^{n} (-1)^{j-i} f_{n-j}(b_j) g(0, b_i, \ldots, b_{j-1}) + f_{n-i}(b_i) = \delta_n^{\,i}$$

and

[2b] $$\sum_{j=i}^{n-1} (-1)^{j-i} f_{j-i}(b_i) g(0, b_j, \ldots, b_{n-1}) + (-1)^{n-i} f_{n-i}(b_i) = \delta_n^{\,i}$$

for all i and n. This fact can be expressed in a concise form: the triangular matrices

$$G_m = \begin{vmatrix} g(0) & g(0,b_1) & g(0,b_1,b_2) & \cdots & g(0,b_1,\ldots,b_m) \\ & g(0) & g(0,b_2) & \cdots & g(0,b_2,\ldots,b_m) \\ & & g(0) & \cdots & g(0,b_3,\ldots,b_m) \\ & & & \cdots & \cdot \\ & & & \cdot\cdot & \cdot \\ & & & \cdot & \cdot \\ 0 & & & & \end{vmatrix}$$

and

$$M_m = \begin{vmatrix} f_0(b_1) & (-1)f_1(b_1) & (-1)^2 f_2(b_1) & \cdots & (-1)^m f_m(b_1) \\ & f_0(b_2) & (-1)f_1(b_2) & \cdots & (-1)^{m-1} f_{m-1}(b_2) \\ & & f_0(b_3) & \cdots & (-1)^{m-2} f_{m-2}(b_3) \\ & & & \cdots & \cdot \\ & & & \cdot\cdot & \cdot \\ & & & \cdot & \cdot \\ 0 & & & & f_0(b_{m+1}) \end{vmatrix}$$

are inverse to one another for any $m \geq 0$. Indeed this property is instrumental in proving [1k]. Moreover, the use of matrix form makes the work simpler. For instance, the inversion formulae corresponding to [2a] and [2b] can be elegantly written as

[2c] $X_m = G_m Y_m$ if and only if $Y_m = M_m X_m$

where $X_m' = (x_0, x_1, \ldots, x_m)$ and $Y_m' = (y_0, y_1, \ldots, y_m)$. As a series relation, [2c] is equivalent to

$$x_n = y_n + \sum_{r=n+1}^{m} g(0, b_{n+1}, \ldots, b_r) y_r$$

if and only if

$$y_n = \sum_{r=n}^{m} (-1)^{r-n} f_{r-n}(b_{n+1}) x_r$$

for any $n \leq m$, $m \geq 0$.

An immediate consequence of the matrix form is that G_m' and M_m' are inverses of each other and therefore

[2d] $X_m = G_m' Y_m$ if and only if $Y_m = M_m' X_m$.

Once again, by specializing $f_i(\xi)$, we can generate several inversion formulae. For example, we may take

$$f_i(\xi) = \frac{\xi}{\xi + id} \begin{bmatrix} \xi + id \\ i \end{bmatrix}$$

and get a new set of formulae.

Put $a_i = 0$ in [1k] and write the identities in matrix form. We immediately observe that G_m and another triangular matrix are inverses of each other. Because of the uniqueness of the inverse, we have derived no new inversion formula. However, the identity can be written in an alternative form which leads to the discussion of quasi-orthogonality as in references 9 and 21; we omit details.

We do not include many special cases; the interested reader may find them in papers included in the references. The multidimensional orthogonal relations and the corresponding inversion formulae are given in reference 20.

3

As was shown by Stanley (reference 26), lattice paths should be in some way associated with trees, and therefore could give rise to identities related to paths. Counting trees in two different ways has given rise to combinatorial identities in reference 13. We illustrate this fact by two examples for which we need some definitions.

Let $T = (V, E, v, \alpha)$ be a planted plane (for short p.p.) tree where V is the vertex set, E the edge set, v the root with degree one, and α an order relation satisfying the following properties:

(i) For $x, y \in V$, if $\rho(x) < \rho(y)$, then $x \,\alpha\, y$, where $\rho(x)$ is the length of the path from v to x.

(ii) If $\{r, s\}$, $\{x, y\} \in E$, $\rho(r) = \rho(x) = \rho(s) - 1 = \rho(y) - 1$ and $r \,\alpha\, x$, then $s \,\alpha\, y$.

Two p.p. trees (V, E_1, v_1, α_1) and (V, E_2, v_2, α_2) are isomorphic if there exists a permutation ϕ on V such that $\phi(v_1) = v_2$, $E_2 = \{\{\phi(x), \phi(y)\} : \{x, y\} \in E_1\}$, and $x \,\alpha_1\, y$ if and only if $\phi(x) \,\alpha_2\, \phi(y)$.

Denote by $P(V, q)$ the set of p.p. trees such that among $n+1$ vertices which have degree greater than one, the ith one (as ordered by α) has exactly degree $q_i + 1$, $i = 1, \ldots, n+1$. Also denote by $P^*(V, k)$, $k = (k_1, \ldots, k_r)$, the set of p.p. trees such that exactly $k_i (\geq 0)$ vertices have degree $i+1$, $i = 1, \ldots, r$. Note that $|V| = \sum_{i=1}^{r} ik_i + 2$. Let IP stand for the set of classes of isomorphic trees in any set of trees P.

Using these definitions, we know from reference 1 that the problem is directly solved by Kreweras's Theorem 1A of Chapter II. Clearly,

[3a] $\quad |IP(V, q)| = \det_{n \times n} \|c_{ij}\| \qquad$ (see [1d])

with $a_i = \sum_{j=1}^{i} q_j - i$, $i = 1, \ldots, n$, and

[3b] $$|IP^*(V, k)| = \frac{1}{r+1} \begin{bmatrix} r+1 \\ k_1, \ldots, k_r \end{bmatrix}$$

where $r = \sum_{i=1}^{r} ik_i = |V| - 2$. These have been derived by using path correspondence of the set of trees under consideration. Suppose that among the q_i's there are exactly k_i q's each equal to i, $i = 1, \ldots, r$. Hence, we get the identity

[3c] $\quad \sum |IP(V, q)| = |IP(V, k)|$

where the summation is over all distinct permutations of $(q_1, \ldots, q_{n+1})$. The proof of the identity through the interpretation of paths does not seem to be straightforward.

Again let $P(V)$ be the set of p.p. trees with $|V|$ vertices. Harris (1952) has used path correspondence to show that

[3d] $$|IP(V)| = \frac{1}{2|V|-1} \begin{bmatrix} 2|V|-1 \\ |V|-2 \end{bmatrix}.$$

Thus we have another identity as follows:

[3e] $$\sum \frac{1}{r+1} \begin{bmatrix} r+1 \\ k_1, \ldots, k_r \end{bmatrix} = \frac{1}{2r+1} \begin{bmatrix} 2r+1 \\ r \end{bmatrix},$$

the summation being over all possible $(k_1, \ldots, k_r)$ which are solutions of $\sum_{i=1}^{r} ik_i = r$.

4
Other identities arise in somewhat different but related situations. The first case is associated with generalized Fibonacci numbers $N(n)$ given by the recurrence relation

[4a] $N(n) = N(n-1)+N(n-r)$

with $N(n) = 0$ for $n<0$, and $N(n) = 1$ for $n = 1, \ldots, r$.

Following Narayana (reference 24), we partition $N(n)$ into $N(n; u_1, \ldots, u_r)$ which satisfy a certain set of recurrence relations (see references 16 and 17). The exact evaluation of $N(n; u_1, \ldots, u_r)$ is done, so that we get an identity. Fibonacci numbers can also be interpreted through paths. Let $f_r(n, y)$ be the number of paths from the origin r to the point $(n-ry, y)$ (r being a positive integer) which is on the boundary $x+ry = n$. Let

$$N(n) = \sum_y f_r(n, y).$$

Then $N(n)$ satisfies [4a].

The second case arises by a close analysis of path enumeration [11] in Section 1. Let $\sigma = (x_1, \ldots, x_n)$ be a sequence of n real numbers and let

$$s_j = \sum_{i=1}^{j} x_i, \quad M_j(\sigma) = \max\{s_1, \ldots, s_j\}, \qquad 1 \leq j \leq n,$$

with $x^+ = \max\{0, x\}$ and $\sigma_i = (x_i, x_{i+1}, \ldots, x_{n+i-1})$ with $x_{n+r} = x_r$, $i = 1, \ldots, n$. Then we have the identity

[4b] $$\sum_{i=1}^{n} (M_n^+(\sigma_i) - M_{n-1}^+(\sigma_i)) = s_n^+$$

due to Dwass (reference 3). The counting result follows as a special case of [4b]. It has been further generalized by several authors (see references 4, 11, 18).

REFERENCES

1 Chorneyko, I. Z., and S. G. Mohanty. 1975. 'On the enumeration of certain sets of planted plane trees,' *J. Comb. Theory 18*, 209–21

2 Dvoretzky, A., and T. H. Motzkin. 1947. 'A problem of arrangements,' *Duke Math. J. 14*, 305–13

3 Dwass, M. 1962. 'A fluctuation theorem for cyclic random variables,' *Ann. Math. Statist. 33*, 1450–4

4 Graham, R. L. 1963. 'A combinatorial theorem for partial sums,' *Ann. Math. Statist. 34*, 1600–2

5 Gould, H. W. 1956. 'Some generalizations of Vandermonde's convolution,' *Amer. Math. Monthly 63*, 84–91

6 – 1957. 'Final analysis of Vandermonde's convolution,' *Amer. Math. Monthly 64*, 409–15

7 – 1960. 'Generalization of a theorem of Jensen concerning convolutions,' *Duke Math. J. 27*, 71–6

8 – 1962. 'A new convolution formula and some new orthogonal relations for inversion series,' *Duke Math. J. 29*, 393–404

9 – 1965. 'The construction of orthogonal and quasiorthogonal number sets,' *Amer. Math. Monthly 72*, 591–602

10 GROSSMAN, H. D. 1950. 'Fun with lattice points,' *Scripta Math. 16*, 119–24

11 HARPER, L. H. 1966. 'A family of combinatorial identities,' *Ann. Math. Statist. 37*, 509–12

12 HARRIS, T. E. 1952. 'First passage and recurrence distributions,' *Trans. Amer. Math. Soc. 73*, 471–86

13 KLARNER, D. A. 1970. 'Correspondences between plane trees and binary sequences,' *J. Comb. Theory 9*, 401–11.

14 KREWERAS, G. 1965. 'Sur une classe de problèmes de dénombrement liés au treillis des partitions des entiers,' *Cahiers du Bur. Univ. de Rech. Opér. 6*, 5–105

15 MOHANTY, S. G. 1966. 'Some convolutions with multinomial coefficients and related probability distributions,' *SIAM Rev. 8*, 501–9

16 – 1967. 'An *r*-coin-tossing game and the associated Fibonacci numbers,' *Sankhya*, Ser. A, *29*, 207–14

17 – 1968. 'On a partition of generalized Fibonacci numbers,' *Fibonacci Quart. 6*, 22–33

18 – 1971. 'A note on combinatorial identities for partial sums,' *Can. Math. Bull. 14*, 65–7

19 – 1972. 'On queues involving batches,' *J. Appl. Prob. 9*, 430–5

20 MOHANTY, S. G., and B. R. HANDA. 1969. 'Extensions of Vandermonde type convolutions with several summations and their applications,' *Can. Math. Bull. 12*, 45–62

21 – 1970. 'A generalized Vandermonde-type convolution and associated inverse series relations,' *Proc. Camb. Phil. Soc. 68*, 459–74

22 MOHANTY, S. G., and T. V. NARAYANA. 1961. 'Some properties of compositions and their application to probability and statistics I,' *Biom. Z. 3*, 252–8

23 NARAYANA, T. V. 1955. 'A combinatorial problem and its application to probability theory I,' *J. Indian Soc. Agric. Statist. 7*, 169–78

24 – 1962. 'An analogue of the multinomial theorem,' *Can. Math. Bull. 5*, 43–50

25 RIORDAN, J. 1968. *Combinatorial Identities.* John Wiley and Sons, New York

26 STANLEY, R. L. 1972. *Ordered Structures and Partitions.* American Mathematical Society Memoirs No. 119

27 TAKÁCS, L. 1962. 'A generalization of the ballot problem and its application in the theory of queues,' *J. Amer. Statist. Assoc. 57*, 327–37

Notes and solutions

References given at the end of each chapter are referred to by chapter (e.g. [III, 2] for reference 2 of Chapter III), and references in the bibiliography at the end of the book are referred to by number only (e.g.[3]). When complete solutions are available, only a brief reference to the literature is made.

CHAPTER I

The central theme of this book is Kreweras's dominance theorem (Theorem 1A, Chapter II) and its application to statistical inference in the case of Lehmann alternatives as initiated by Steck (Equation [3e], Chapter II). Historically the fundamental problem of Young chains or the number of standard Young tableaux has been solved by many authors. MacMahon, in his *Combinatory Analysis* (1915), posed and solved it in the following form:

Y1: In a ballot problem involving k candidates, a well-informed observer knows beforehand that the candidates will obtain $y_1, \ldots, y_k$ votes where $y_1 + \ldots + y_k = n$ and $y_1 \geq y_2 \geq \ldots \geq y_k > 0$. Assuming that the $n!/y_1 \times \ldots \times y_k$ possible orders of votes are equally likely, what is the probability that at every stage of the counting the votes $c_1, \ldots, c_k$ obtained by candidates $1, \ldots, k$ satisfy $c_1 \geq \ldots c_k \geq 0$?

Although the statement of this ballot problem is similar to André's problem when $k = 2$, the methods used to solve Y1 are very different when $k > 2$. Simultaneously, in connection with the representation theory of the symmetric group S_n, Frobenius and Young obtained the equivalent result:

Y2: Let $Y = (y_1, \ldots, y_k)$ be a partition of n, i.e. $y_1 \geq y_2 \geq \ldots \geq y_k > 0$ and $W(y) = \sum y_i = n$. To each irreducible representation of S_n corresponds a partition of n $(y_1, \ldots, y_k)$, and its degree is given by

$$f(Y) = n! \frac{\prod_{i<j}(y_i - y_j - i + j)}{\prod_i (y_i + k - i)!}.$$

It is now usual to refer to the above solution as Young's formula and to treat Y1 and Y2 as brickwall profiles as follows. Let us assume for aesthetic convenience (and not as a mathematical restriction) that bricks are laid or votes tallied at the uniform rate of 1 per second.

Y: What is the probability, in building a brickwall with final column heights $(y_1, \ldots, y_k)$, that at every instant $t (0 < t \leq n)$ the column heights $c_1(t), \ldots, c_k(t)$ satisfy $c_1 \geq \ldots c_k \geq 0$?

The brickwall profile regarded geometrically as in Figure 7 immediately gives rise to the elegant *théorème d'équerre* (Hook Theorem) of Frame-Robinson-Thrall (1954):

R: $f(Y) = n!/\Pi_{i,j} h_{ij}$ where h_{ij} are the lengths of the hooks in the Young tableau (exercise 10, Chapter II).

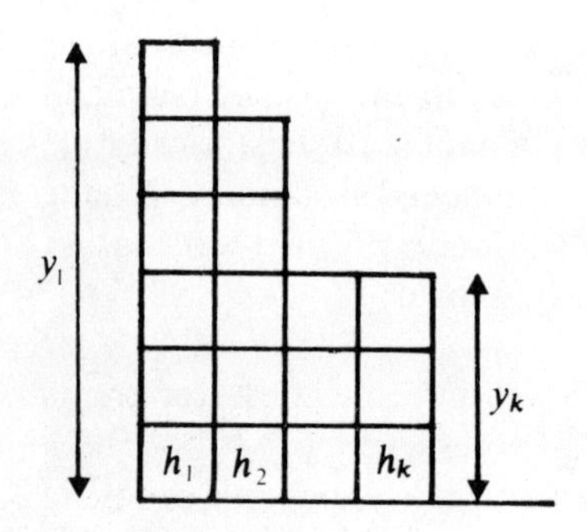

Figure 7. A 'brickwall.'
(The hooks of the lowest row are given by $h_i = y_i + k - i$.)

The above result Y or problem of Young was generalized by Kreweras [II, 3], and we adopt the following terminology, which is most suitable for applications. Let the lattice of partitions under domination (cf. Narayana [III, 12], Kreweras [II, 3], and Berge [1]) be called the Young lattice. This lattice is obviously distributive, and if $A = (a_1, \ldots, a_k)$ dominates $B = (b_1, \ldots, b_k)$ and $Y(A, B)$ represents the set of *Young chains* in going from B to A in the lattice, the enumeration of this number of Young chains is achieved by Y1, Y2, or R when $B = \mathbf{0}$ (vector with all elements zero). Kreweras achieves the enumeration of these Young chains in the general case when $B \neq \mathbf{0}$. Letting $Y(A, B)$ be the number of Young chains from B to A, then

K-Y: $|Y(A,B)| = [W(A) - W(B)]! \| [(a_i - i) - (b_j - j)] \# ! \|$

where the (i,j) element of the $k \times k$ determinant is given on the right-hand side and $z\#!$ is 0, 1, or $(z!)^{-1}$ according as z is negative, zero, or a positive integer. It is a simple exercise to state K-Y1, which generalizes Y1 (exactly as K-Y generalizes Y), and Kreweras [II, 3] has shown that when $B \neq \mathbf{0}$ no simple expression for 'hooks' seems to exist corresponding to R. However, when $B = \mathbf{0}$, Kreweras shows the equivalence of Y with R, by a very natural argument.

Brief comments on exercises

1, 2, 3. [15].

4. This result formed part of the folklore of at least the 1950s – if not earlier – and came to my notice through a referee. A proof is in [9]. However, the fact that the result can be derived effortlessly from a one-sided dominance theorem similar to Theorem 1A of Chapter II escaped the attention of mathematicians for at least two decades.

7. Dominations were first introduced for compositions (see [III, 12]) in 1955, and the equivalence of various forms of the definition as applied to ranks in statistics, distributing balls into boxes, or dominance in the Young lattice was realized only slowly. As one instance, the statement of exercise 7(d), published in 1958, was rediscovered in rank form by Steck in 1974 (see [II, 7]); Steck handsomely acknowledged priority. Again the original dominance lattice of compositions of an integer reappears as Kreweras's Young lattice a decade later (see [II, 3]). Such instances can be multiplied.

9. (a) For a solution in English see [I, 2].
(c) A typical Leo Moser solution; more generally see [12].

10. [I, 9]; see also [4], [26].

11. [7].

CHAPTER II

The dominance theorem consists of two parts, which have been developed at the start of Chapters II and III. The first part involves the now familiar (Kreweras) numbers $K_r(A,B)$ as defined in equation [1c], Chapter II. It is these numbers, or more specifically these numbers for $r = 1$, which have been repeatedly rediscovered and interpreted in slightly different forms by Steck, Mohanty, Pitman [21], Sarkadi [26], and perhaps others. However, the second part of the dominance theorem yields the identity connecting $|Y_r(A,B)|$ and K_r (equation [1b], Chapter II). To enable the reader to complete the proof, the following details are given. Starting from the set $Y(A,B)$ of p. 48, i.e.

The Set Y(A,B)

$$\begin{array}{cccccccc}
 & & & 322 & & & & \\
321 & 321 & 321 & 321 & 321 & 321 & \underline{222} & \underline{222} \\
\underline{221} & 320 & 320 & \underline{311} & 311 & \underline{221} & 221 & 221 \\
220 & \underline{220} & 310 & 310 & \underline{211} & \underline{211} & \underline{211} & 220 \\
 & & & 210 & & & &
\end{array}$$

let us consider the set $Y^*(A,B)$ which consists of $Y(A,B)$ 'thinned out,' i.e., where we retain, apart from the end-points, only the switchback points.

The Set Y(A,B)*

$$\begin{array}{cccccccc}
322 & 322 & 322 & 322 & 322 & 322 & 322 & 322 \\
 & & & \underline{311} & & & \underline{221} & \underline{222} \\
 & \underline{221} & \underline{220} & & \underline{211} & \underline{222} & \underline{211} & \underline{211} \\
210 & 210 & 210 & 210 & 210 & 210 & 210 & 210
\end{array}$$

Now every element of $K_n(A,B)$ when 'thinned out' (i.e. by (a) removing all repetitions and (b) removing all non-switchback points) reduces to an element of $Y^*(A,B)$. Conversely, let us consider a particular chain in $Y(A,B)$ with r switchbacks, say $Y_r^0(A,B)$. The corresponding element in the set $Y^*(A,B)$ has $r+2$ points or has length $r+2$. Placing $n-r$ balls in the $W(A)-W(B)+1$ boxes represented by the elements of $Y_r^0(A,B)$ itself, we obtain all elements of $K_n(A,B)$ which can be generated from $Y_r^0(A,B)$. Therefore

$$|K_n(A,B)| = \sum_{r=0}^{n} |Y_r(A,B)| \binom{n-r+w(A)-W(B)}{n-r}.$$

Concluding our example, what is the contribution of $Y_1^0(A,B) = [210\ 220\ \underline{221}\ 321\ 322]$ to $|K_3(A,B)|$? Place two balls in the five boxes given by points of $Y_1^0(A,B)$. If we choose the particular placement $(0,0,1,1,0)$, say, the corresponding sequence in $K_3(A,B)$ is $210\ 221\dagger\ \underline{221}\ 321\dagger\ 322$, where elements with daggers (†) are precisely the additions to $[210\ \underline{221}\ 322] \in Y^*(A,B)$ and are specified by $(0,0,1,1,0)$. Thus $Y_1^0(A,B)$ contributes ${}^6C_2 = 15$ cases to $|K_3(A,B)| = 120$.

As further illustrations of the dominance theorem and for the sake of completeness, we include tables of switchbacks for all partitions of $n\,(6 \le n \le 10)$ due to McKay and Rohlicek and also tables of switchbacks over the $m \times n$ rectangle due to Aiello (Tables 11 and 12). The latter tables are for $6 \le m+n \le 11$ with the further conditions $\min(m,n) \ge 3$, $\max(m,n) \le 7$. The case where $\min(m,n) = 2$ yields the Runyon or Narayana numbers, which have been thoroughly studied by Kreweras and Narayana (see Chapter V).

Finally, some comment about the various types of duality (Definitions A and B and Theorem 3A, chapter II) seems in order. We refer to the admirable work by Milton [8]. Here tables of probability duality under shift alternatives for normal

TABLE 11

Tables of switchbacks
Let P be a partition of n. Then we enter the number of cases of N switchbacks under 1, 2, …. Under N the *sum* of all switchbacks is entered. This is the degree of the irreducible representation of S_n corresponding to the partition of n. We omit the trivial partition of n (n itself) and the column of 1's under 0. Conjugate partitions are also omitted.

P				N	1	2	3	4	5	6
$n = 6$										
5	1			5	4					
4	2			9	5	3				
4	1	1		10	6	3				
3	3			5	3	1				
3	2	1		16	7	7	1			
$n = 7$										
6	1			6	5					
5	2			14	7	6				
5	1	1		15	8	6				
4	3			14	6	6	1			
4	2	1		35	11	18	5			
4	1	1	1	20	9	9	1			
3	3	1		21	8	10	2			
$n = 8$										
7	1			7	6					
6	2			20	9	10				
6	1	1		21	10	10				
5	3			28	9	14	4			
5	2	1		64	15	34	14			
5	1	1	1	35	12	18	4			
4	4			14	6	6	1			
4	3	1		70	14	34	19	2		
4	2	2		56	13	28	13	1		
4	2	1	1	90	17	45	25	2		
3	3	2		42	10	20	10	1		
$n = 9$										
8	1			8	7					
7	2			27	11	15				
7	1	1		28	12	15				
6	3			48	12	25	10			
6	2	1		105	19	55	30			
6	1	1	1	56	15	30	10			
5	4			42	10	20	10	1		
5	3	1		162	20	70	60	11		
5	2	2		120	18	55	40	6		

P					N	1	2	3	4	5	6
5	2	1	1		189	23	84	70	11		
5	1	1	1	1	70	16	36	16	1		
4	4	1			84	15	40	25	3		
4	3	2			168	18	65	65	18	1	
4	3	1	1		216	22	85	85	22	1	
3	3	3			42	10	20	10	1		
$n = 10$											
9	1				9	8					
8	2				35	13	21				
8	1	1			36	14	21				
7	3				75	15	39	20			
7	2	1			160	23	81	55			
7	1	1	1		84	18	45	20			
6	4				90	14	40	30	5		
6	3	1			315	26	118	135	35		
6	2	2			225	23	91	90	20		
6	2	1	1		350	29	135	150	35		
6	1	1	1	1	126	20	60	40	5		
5	5				42	10	20	10	1		
5	4	1			288	23	100	120	41	3	
5	3	2			450	26	133	195	86	9	
5	3	1	1		567	31	169	250	106	10	
5	2	2	1		525	30	159	230	96	9	
5	2	1	1	1	448	31	156	196	61	3	
4	4	2			252	20	85	105	38	3	
4	4	1	1		300	24	105	125	42	3	
4	3	3			210	19	75	85	28	2	
4	3	2	1		768	31	187	330	187	31	1

populations are given – the most important case in applied statistics. Unlike the probability distributions induced by Lehmann alternatives, dual paths have the same probabilities for normal shift alternatives when $m = n$. Also, if $A\,D\,B$, Savage [22] showed that $P_S(B) \geq P_S(A)$, exactly as proved in Lemma 2A, Chapter III, for Lehmann alternatives. Here P_S or $P_{S,d}$ refers to the probability distribution induced by a shift d for normal populations, where d is the ratio of the difference of means and the common standard deviation of the normal populations. Thus although Lehmann alternatives have a duality between k and $1/k$ (equation [3b], Chapter II), normal shift alternatives have path duality. Indeed, Milton [8, p. 3, equation (3.1)] quotes a result of I. R. Savage which implies the first sentence of Lemma 3A, Chapter II. Although path duality is well known, lattice duality was first stated in 1958 (see [19]). Finally exercise 12, which is probably well known, proves – under the null hypothesis – normality for areas under self-dual paths in a

TABLE 12

Table of Young chains $Y_r(m, n)$ on an $m \times n$ rectangle with r switchbacks (Note that the symmetric table is extended as illustrated* and $Y_0 = 1$)

(m, n)	Y_1	Y_2	Y_3	Y_4	Y_5	Y_6	Y_7
(3, 3)	10	20	(10)	(1)			
(4, 3)	22	113	190	(113)			
(5, 3)	40	400	1,456	2,212	(1,456)		
(4, 4)	53	710	3,548	7,700	(7,700)		
(6, 3)	65	1,095	7,095	20,760	29,484		
(5, 4)	105	2,856	30,422	151,389	385,029	523,200	
(6, 4)	185	8,910	171,580	1,596,770	7,962,636	22,599,115	37,775,705
(5, 5)	226	13,177	306,604	3,457,558	21,034,936	73,605,961	153,939,214
	$Y_8 = 196{,}433{,}666$						
(6, 5)	431	47,127	2,057,279	44,065,213	520,591,836	3,650,698,608	15,947,603,944
	$Y_8 = 44{,}808{,}819{,}900$		$Y_9 = 82{,}654{,}311{,}948$		$Y_{10} = 101{,}243{,}378{,}236$		

* Values where m and n are both even are symmetric, the underlined value repeating as when $m = n = 4$.

combinatorial setting. Under very general conditions, the Chernoff-Savage Theorem (1958) – see [2] – proves normality for all paths even under the alternative hypothesis by choosing a very general score function (which includes areas under paths). A combinatorial proof of probability duality under Lehmann alternatives, or even of the Lehmann distribution itself, is unknown to me, although it would be of exceptional interest.

Brief comments on exercises

1. [21].
3–6. [II, 2 and 8].
7. [V, 6].
8–10. [III, 7] is a general reference. A proof of 9(c) is announced in [18].
12. [13].

CHAPTER III

Professor I. R. Savage has very kindly pointed out to me that in Chapter III, Section 2.2, my definition of the UMP test at every level α might be better designated as MPRT or most powerful rank test. I agree that such a designation removes all possibility of falsely interpreting the word *uniformly* as *uniformly in all k*. Apart from Professor Savage's important statistical work on rank-orders (references [27], [28], [29], which are a must for all statisticians if they wish to appreciate any differences in terminology), I also draw attention to [20], which shows the inadmissibility of the Wilcoxon test – so dear to statisticians – against Lehmann alternatives.

I include, in lieu of acknowledgment of priority, a short comment about Lemma 2D, Chapter III.

The following anecdote in connection with simple sampling plans (s.s.ps.) is perhaps amusing and revealing. In 1959, I had worked out one morning that 1, 3, 12 were the number of s.s.ps. for $n = 1, 2, 3$ and, over lunch, mentioned casually to (the late) Leo Moser that I was checking my list of fifty or so plans in the afternoon for $n = 4$. He instantly remarked that this number must be 55 and, in general, was $\frac{1}{n}\binom{3n}{n-1}$ if it were concerned with lattice paths. The late 'grand' Yu. V. Linnik, who expressed to me his high appreciation of Moser's ingenuity in combinatorics during our day's meeting in Paris in 1967, would have been delighted with this story. At that time I had no notion that Linnik would put DeGroot's Theorem in the subject index of his forthcoming book with Professors Kagan and Rao. His co-authors know this story and have received a full account of Barron Brainerd's elegant prior contribution.

A formal definition of a sampling plan will be given following DeGroot [III, 2] for completeness. However, the essentially combinatorial nature of the problem was totally overlooked, and perhaps still is misunderstood, despite Kagan, Linnik, and Rao. The solution is pure 'Young' as was first seen in 1959.

It is helpful to visualize a sequential sample as a minimal lattice path in the Euclidean plane – the path starts at the origin and is extended at a given stage by one unit in either the horizontal or vertical direction. The path stops when a boundary point is reached – equivalently the particle performing the random walk is absorbed. A *sampling plan* (s.p.) S is a function defined on the lattice points γ of the Euclidean plane, i.e. the coordinates $X(\gamma)$, $Y(\gamma)$ are non-negative integers. Furthermore, S takes only the values 0 and 1, and is such that $S(\text{origin}) = 1$. Under a given s.p. γ is a *boundary point* (b.p.) if and only if there exists a path to γ and $S(\gamma) = 0$. It is a *continuation point* (c.p.) if and only if there is a path to γ and $S(\gamma) = 1$, so that there is at least one lattice path through γ. A point γ is an *inaccessible point* (i.p.) if and only if there does not exist any path to γ. Thus every lattice point can be uniquely classified as a c.p., i.p., or b.p. The *sample size* $N(\gamma)$ of a point is the sum of its coordinates, $X(\gamma) + Y(\gamma)$. The *boundary* B is the set of all b.ps. The probability of reaching a particular b.p. γ is $K(\gamma)p^{X(\gamma)}q^{Y(\gamma)}$ where, as usual, $p+q = 1$ with $0<p, q<1$, so that the probability of moving horizontally is p, and $K(\gamma)$ is the number of distinct paths to γ. A s.p. is closed if $\sum_{\gamma \in B} K(\gamma)p^{X(\gamma)}q^{Y(\gamma)} = 1$ for all $0<p<1$. A s.p. is *simple* if and only if for each positive integer m, the c.ps. of sample size m form an interval on the line $X(\gamma) + Y(\gamma) = m$. This interval may be empty or consist of a single point. The functions defined on B and taking the values $X(\gamma)$, $Y(\gamma)$, and $N(\gamma)$ are denoted by X, Y, and N, respectively. Finally a s.p. is *bounded* if and only if there exists a positive integer n such that $P\{n \geq N|p\} = 1$. The *size* of a s.p. which is bounded is the smallest n for which the last relation holds. The problem is to enumerate all simple s.ps. or equivalently all s.ps. of size n with exactly $n+1$ b.ps.

Brief comments on exercises

2–6. A solution to these problems is contained (essentially) in a recent paper by Narayana, Rao, and Pandya (Cahiers du Bier, Univ. de Rech. Opér. 26 (1977), 51–65).

11, 12. [14].

13, 6. Show the weak inadmissibility and inadmissibility of the Wilcoxon test in small samples for normal shift and Lehmann alternatives respectively. It seems not to be recognized by statisticians that the level of a test may influence its (weak) inadmissibility. Thus the weakly inadmissible Wilcoxon test continues to be used for small samples. Of course, the Fisher-Yates-Terry-Hoeffding test is conjectured to be strongly admissible in the class $\{c_1, w, DR(a)\}$.

CHAPTER IV

Most of the exercises are from my 'Contributions to the Theory of Tournaments' (1969), lecture notes given at the University of Alberta. There is a close connection between random tournaments as developed in exercises 4, 5, 7, 8 and Hartigan's deterministic tournaments in [3]. For a compact development of most tournament results we refer to the lecture notes mentioned above, where a simplification of Hartigan's result is achieved.

Yet another amusing connection between tournaments and dominance is the following problem from [31]. The problem here is to assign integer weights to n directors on a board so that (i) different subsets have different total weights (no ties possible in voting) and (ii) every subset of size x has always more weight than every subset of size $x-1$ ($x = 1, \ldots, n$). The subsets, satisfying (ii), may be said to be non-distorting in that every majority beats a minority. A solution to this problem is given in Table 13; the I_m sequence along the main diagonal coincides with the T_n sequence of exercise 1c, Chapter IV. Very great difficulties appear to be encountered when we want to show that such a solution is minimal-sum; indeed, it has neither been proved nor disproved that this solution is minimal-dominance, a property weaker than minimal-sum. Kreweras (personal communication) has discovered some pretty properties of this Narayana-Zidek-Capell sequence.

TABLE 13

Table of non-distorting, tie-avoiding integer vote weights W_m

Members, m	1	2	3	4	5	6	7	8	9	10	11	12	13
Totals, S_m	1	3	9	21	51	117	271	607	1363	3013	6643	14491	31495
Column	1	2	4	7	13	24	46	88	172	337	667	1321	2629
vectors of vote		1	3	6	12	23	45	87	171	336	666	1320	2628
weights, $[W_m]$			2	5	11	22	44	86	170	335	665	1319	2627
				3	9	20	42	84	168	333	663	1317	2625
					6	17	39	81	165	330	660	1314	2622
						11	33	75	159	324	654	1308	2616
							22	64	148	313	643	1297	2605
								42	126	291	621	1275	2583
									84	249	579	1233	2541
										165	495	1149	2457
											330	984	2292
												654	1962
													1308

NOTE: The underlined values along the diagonal of vector elements are the I_m values, where $I_1 = I_2 = 1$ and $I_m = 2I_{m-1} - [\mathrm{mod}_2(m-1)]I_{[m/2]-1}$ for $m \geq 3$. Thus the I_m sequence counts the number of random knock-out tournaments on n players; see [IV, 1 and 9].

CHAPTER V

The example discussed in Chapter III, Section 2.1, is not as artificial as it first appeared!

I am very grateful to Professor Germain Kreweras who, during his delightful but all too short visit to Edmonton and California during summer 1976, informed me that the numbers (n, r) are known as the Narayana numbers in France. Following his kind suggestions, I conclude with a table of these numbers (the well-known Catalan numbers are in the last column) and an asymptotic expansion due to my colleague Professor J. R. McGregor, which yields yet another but very elegant 'asymptotic normality proof' for the Narayana numbers.

The asymptotic normality of (n, r) might be inferred by other methods, such as by the theory of runs or from their factorial moments as given by Kreweras:

$$\sum_{s=0}^{n} (s)_k \binom{n}{s} \binom{n+1}{s} \Big/ (s+1) = \frac{(n)_k (n+1)_k}{(2n+2)_k} C_{n+1}.$$

For an asymptotic expansion of the 'Narayana' polynomials, McGregor proceeds as follows:

The numbers (n, r)

n \ r	1	2	3	4	5	6	7	Totals C_n
1	1							1
2	1	1						2
3	1	3	1					5
4	1	6	6	1				14
5	1	10	20	10	1			42
6	1	15	50	50	15	1		132
7	1	21	105	175	105	21	1	429

It is readily verified that the distribution

$$p_n(r) = \left[\frac{n}{n+1}\binom{2n}{n}\right]^{-1}\binom{n}{r}\binom{n}{r-1} \qquad (r = 1, 2, \ldots, n)$$

of the Narayana numbers has mean $(n+1)/2$ and variance $nK_n{}^2/8$ where $K_n{}^2 = (n^2-1)/(n^2-n/2)$. Thus the moment-generating function of the corresponding standardized distribution is

$$m_n(t) = \left[\frac{n}{n+1}\binom{2n}{n}\right]^{-1} \exp\left[-2^{1/2}(n+1)t/n^{1/2}K_n\right]$$

$$\times \sum_{r=1}^{n} \left[\exp(2.2^{1/2}t/n^{1/2}K_n)^r\right]\binom{n}{r}\binom{n}{r-1}.$$

Consider the (Narayana) polynomials

$$N_n(y) = \sum_{r=1}^{n}\binom{n}{r}\binom{n}{r-1}y^r.$$

Making the substitution $y = (x-1)/(x+1)$ in the formula

$$P_n(x) = \sum_{r=0}^{n}\binom{n}{r}\binom{n}{r}\left(\frac{x-1}{2}\right)^r\left(\frac{x+1}{2}\right)^{n-r}$$

for the Legendre polynomial of degree n (Szegö [30]), we obtain, after considerable manipulation,

$$N_n(y) = -\left\{(1-y)^n P_n\left(\frac{1+y}{1-y}\right) + (n+1)y^{n+1}\int \frac{(1-y)^n}{y^{n+2}} P_n\left(\frac{1+y}{1-y}\right)dy\right\}.$$

Using the Laplace-Heine asymptotic formula for the Legendre polynomials [30, p. 194] for $y>0$, we obtain

$$N_n(y) \cong -\frac{1}{2(\pi n)^{1/2}}\left\{(1+y^{1/2})^{2n+1}y^{-1/4}+(n+1)y^{n+1}\int y^{-1/4}(1+1/y^{1/2})^{2n+1}\frac{dy}{y^{3/2}}\right\}.$$

Repeated integration by parts of $\int[y^{-1/4}(1+1/y^{1/2})^{2n+1}dy/y^{3/2}]$ yields the asymptotic expansion

$$-(n+1)^{-1}y^{-1/4}(1+1/y^{1/2})^{2n+2}+\sum_{k=1}^{\infty} a_k\frac{2^{k+1}y^{(2k-1)/4}(1+1/y^{1/2})^{2n+k+2}}{(2n+2)(2n+3)\ldots(2n+k+2)},$$

where $a_1 = 1/4$, $a_k = 1\cdot 3\cdot 5\cdots(2k-3)/4^k$ $(k = 2, 3,\ldots)$. Thus

$$\int y^{-1/4}(1+1/y^{1/2})^{2n+1}\frac{dy}{y^{3/2}} = -(n+1)^{-1}y^{-1/4}(1+1/y^{1/2})^{2n+2}\{1+O(n^{-1})\},$$

so that

$$N_n(y) \cong \frac{1}{2(\pi n)^{1/2}}y^{1/4}(1+y^{1/2})^{2n+1}$$

and

$$m_n(t) \cong \left[\frac{n}{n+1}\binom{2n}{n}\right]^{-1}\frac{2^{2n}}{(\pi n)^{1/2}}[\cosh(t/(2n)^{1/2}K_n)]^{2n+1}.$$

Hence

$$\operatorname*{limit}_{n\to\infty} m_n(t) = e^{t^2/2}$$

which is the moment-generating function of the standardized normal distribution.

Supplementary bibliography

This bibliography supplements the references at the ends of the chapters and is meant to be complete only from the point of view of dominance.

1 Berge, C. 1970. *Principles of Combinatorics*. Academic Press, New York (French version, 1968)

2 Chernoff, H., and I. R. Savage. 1958. 'Asymptotic normality and efficiency of certain nonparametric statistics,' *Ann. Math. Statist. 29*, 972–94

3 Hartigan, J. 1966. 'Probabilistic completion of a knock-out tournament,' *Ann. Math. Statist. 37*, 495–503

4 Hodges, J. L. 1955. 'Galton's rank-order test,' *Biometrika 42*, 261–2

5 Kreweras, G. 1972. 'Sur les partitions non croisée d'un cycle,' *Discrete Math. 1*, 333–50

6 – 1972. 'Classification des permutations suivant certaines propriétés ordinales de leur représentation plane,' *Actes du Colloquium 'Permutations,' CNRS Paris*, 97–115

7 – 1976. 'Aires des chemins surdiagonaux et application à un problème économique,' *Cahiers du Bur. Univ. de Rech. Opér. 24*, 1–8

8 Milton, R. 1970. *Rank Order Probabilities*. John Wiley and Sons, New York

9 Mohanty, S. G. 1961. 'Some properties of compositions and their application to probability and statistics.' Doctoral thesis, University of Alberta

10 – 1971. 'A short proof of Steck's result on two-sample Smirnov statistics,' *Ann. Math. Statist. 42*, 413–14. Acknowledgment of priority, *Ann. Statist. 5* (1977), 429

11 Mohanty, S. G., and T. V. Narayana. 1961. 'Some properties of compositions and their applications to probability and statistics I,' *Biom. Z. 3*, 252–8

12 Moser, L., and W. Zayachkowski. 1961. 'Lattice paths with diagonal steps,' *Scripta Math. 26*, 223–9

13 Mureika, R. A., T. V. Narayana, and M. S. Rao. 1975. 'Areas, dominance and duality,' *Carleton University* (Ottawa) *Statistics Conference Proceedings*, 10.01–10.15

14 Narayana, T. V. 1962. 'An analogue of the multinomial theorem,' *Can. Math. Bull. 5*, 43–50

15 – 1968. 'Cyclic permutations of lattice paths and the Chung-Feller theorem,' *Skand. Aktuarietidskrift* 23–30

16 – (assisted by E. Goodman). 1969. 'A note on a double series expansion,' *Cahiers du Bur. Univ. de Rech. Opér. 13*, 19–24

17 – 1975. 'Sur la théorie de domination et son application aux inférences statistiques,' *Compt. Rend. Acad. Sci. Paris 280*, 25–7

18 – 1975. 'Chaînes de Young et tests non-paramétriques,' *Compt. Rend. Acad. Sci. Paris 281*, 1075

19 NARAYANA, T. V., and G. E. FULTON. 1958. 'A note on the compositions of an integer,' *Can. Math. Bull. 1*, 169–73

20 NARAYANA, T. V., I. R. SAVAGE, and K. M. LAL SAXENA. 197?. 'Young chains and rank orders,' *Can. J. Statist.* (in press)

21 PITMAN, E. J. G. 1972. 'Simple proofs of Steck's determinantal expressions for probabilities in the Kolmogorov and Smirnov tests,' *Bull. Aust. Math. Soc. 7*, 227–32

22 RIORDAN, J. 1958. *An Introduction to Combinatory Analysis.* John Wiley and Sons, New York

23 – 1968. *Combinatorial Identities.* John Wiley and Sons, New York

24 ROBINSON, G. DE B. 1961. *Representation Theory of the Symmetric Group.* University of Toronto Press, Toronto

25 RONDE, TABLE. 1976. *Actes de la Table Ronde du C.N.R.S. du 26–30 avril, Strasbourg* (ed. D. Foata). Springer Verlag, Berlin

26 SARKADI, K. 1961. 'On Galton's rank order test,' *Publ. Math. Inst. Hungar. Acad. Sci. 6*, 127–31. Addendum, ibid. *7* (1962), 223

27 SAVAGE, I. R. 1956. 'Contributions to the theory of rank order statistics: the two-sample case,' *Ann. Math. Statist. 27*, 590–615

28 – 1957. 'Contributions to the theory of rank order statistics: the "trend" case,' *Ann. Math. Statist. 28*, 968–77

29 – 1959. 'Contributions to the theory of rank order statistics: the one-sample case,' *Ann. Math. Statist. 30*, 1018–23

30 SZËGO, G. 1975. *Orthogonal Polynomials.* American Mathematical Society, Colloquium Publications, Providence, R.I.

31 WYNNE, B. E., and NARAYANA, T. V. 197?. 'Tournament configuration and weighted voting,' *Cahiers du Bur. Univ. de Rech. Opér.* (in press)

Index